A Reference Manual for Biomedical Engineering and its Applications

A Reference Manual for Biomedical Engineering and its Applications

Edited by **Mark Walters**

LANRYE INTERNATIONAL

New Jersey

Published by Clanrye International,
55 Van Reypen Street,
Jersey City, NJ 07306, USA
www.clanryeinternational.com

A Reference Manual for Biomedical Engineering and its Applications
Edited by Mark Walters

© 2015 Clanrye International

International Standard Book Number: 978-1-63240-000-0 (Hardback)

Contents

Preface

This book is a comprehensive study of the field of biomedical engineering that is authoritative and up-to-date. It talks about various aspects of biomedical engineering and its applications in science and industry. In this book, there are diverse topics that have been covered, like safety of patients pertaining to medical technology management, use of optics in biomedical technology, methods of skin welding procedures, medical instrument applications and their fundamentals.

After months of intensive research and writing, this book is the end result of all who devoted their time and efforts in the initiation and progress of this book. It will surely be a source of reference in enhancing the required knowledge of the new developments in the area. During the course of developing this book, certain measures such as accuracy, authenticity and research focused analytical studies were given preference in order to produce a comprehensive book in the area of study.

This book would not have been possible without the efforts of the authors and the publisher. I extend my sincere thanks to them. Secondly, I express my gratitude to my family and well-wishers. And most importantly, I thank my students for constantly expressing their willingness and curiosity in enhancing their knowledge in the field, which encourages me to take up further research projects for the advancement of the area.

Editor

Biomaterials

A. Binnaz Hazar Yoruç[1] and B. Cem Şener[2]
[1]Yıldız Technical University, Science and Technology Application and Research Center,
[2]Marmara University, Faculty of Dentistry,
Department of Oral and Maxillofacial Surgery,
Turkey

1. Introduction

Biomaterial can be described as a combination of substances originating from natural, inorganic or organic materials that is biocompatible in exactly or partially contact with the body for healing time. They involve complete or part of a living organism or biomedical device which performs, augments or replacements any natural function [1].

Biomaterial is a nonviable substance used in a medical device intended to interact with biological systems. Their usage within a physiologic medium needs the characteristic features such as efficient and reliable. These characteristic features have provided with a suitable combination of chemical, mechanical, physical and biological properties [2]. Nowadays, biomaterials are commonly used in various medical devices and systems; synthetic skin; drug delivery systems; tissue cultures; hybrid organs; synthetic blood vessels; artificial hearts; cardiac pacemakers; screws, plates, wires and pins for bone treatments; total artificial joint implants; skull reconstruction; dental and maxillofacial applications [3].

Metals and their alloys, polymers, ceramics and composites are commonly used for biomedical applications. These materials that have different atomic arrangement present the diversified structural, physical, chemical and mechanical properties and so different properties offer alternative applications in the body. The mechanical properties of metals and their alloys such as strength, elasticity coefficient and fatigue life makes them attractive materials for many load-bearing biomedical systems. Metallic materials tend to degradation in a corrosion process and even as the corrosion reactions of releasing some side products such as ions, chemical compounds and insoluble components that may cause adverse biological reactions.

Ceramic materials are desirable biomaterials due to the biocompatible properties such as bioactive, bioinert and biodegradable, however they have significant disadvantages such as brittleness, low strength etc. [1-4].

Polymers are attractive materials for biomedical applications such as cardiovascular devices, replacement and proliferation of various soft tissues. They are also used in drug delivery systems, diagnostic supports and as a reconstructive material for tissue engineering. The current applications of them include cardiac valves, artificial hearts, vascular grafts, breast prosthesis, contact and intraocular lenses, fixtures of extracorporeal oxygenators, dialysis and plasmapheresis systems, coating materials for medical products, surgical materials, tissue adhesives etc. [3]. The composition, structure and organization of constituent macromolecules specify the properties of polymers.

Composite is a material comprised of two or more metal, polymer or ceramic structures which are separated by an interface. Composite materials have been widely used for a long time in innovative technological applications due to their superior mechanical properties. The bones, tendons, skin, ligaments, teeth, etc. are natural composite structures in the human body. The amount, distribution, morphology and properties of structure components determine the final behavior of resultant tissues or organs. Some synthetic composites can be used to produce prosthesis able to simulate the tissues, to compromise with their mechanical behavior and to restore the functions of the damaged tissues. Composites are usually classified based on their matrix components like metals, ceramics, polymers or reinforcement components like particulates, short or long fibers, microfillers, nanofillers. Many matrix and reinforcement components of composite materials have been tried by several researchers in tissue engineering to advance the mechanical features, biologic functions and to deliver special molecules. Biocompatible polymers have been mostly applied as matrix for composite materials associated with ceramic fillers in tissue engineering. Although ceramics are generally stiff and brittle materials, polymers are known to be flexible and exhibit low mechanical strength and stiffness. Composites aim to combine the properties of both materials for medical applications [1-9].

There are currently thousands of surgical materials, hard and soft tissue products, biomedical devices, pharmaceutical and diagnostic products and disposable materials at the medical market. The recent biomaterial applications were aimed to engineered tissues, intelligent materials, tissue cultures, drug delivery systems, artificial organs, biomimetic systems and materials in addition to traditional medical applications. Nowadays, modern clinical procedures such as preventing and curing main genetic diseases are become significant and new medical demands cause the change of the biomaterial products. Materials scientists and engineers need researchers who work effectively in professional teams such as molecular biologists, biochemists, geneticists and physicians and they also aim for the materials which are recognised by cells, biochemical structures, molecules and genetic issues [4].

2. Materials used in medicine

2.1 Natural materials

Biopolymers are natural materials such as carbohydrates, proteins, cellulose, starch, chitin, proteins, peptides, DNA and RNA produced by living organisms [10]. Generally, the synthesis procedure of them involves enzyme-catalysed polymerization reactions of activated monomers and chain growth which are characteristically formed within cells by complex metabolic processes [11]. Figure 1 shows classification, structures and functions of biopolymers. Table 1 also presents incidence and physiological functions of certain natural polymers.

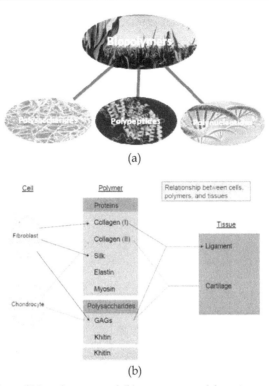

(a)

(b)

Fig. 1. (a) Classification of biopolymers and (b) structure and functions of biopolymers [12].

	Polymer	Incidence	Physiological Function
Proteins	Silk	Synthesized by arthropods	Protective cocoon
	Keratin	Hair	Thermal insulation
	Collagen	Connective tissues (tendon, skin, etc.)	Mechanical support
	Gelatine	Partly amorphous collagen	
	Fibrinogen	Blood	Blood clotting
	Elastin	Neck ligament	Mechanical support
	Actin	Muscle	Contraction, motility
	Myosin	Muscle	Contraction, motility
Polysaccharides	Cellulose (cotton)	Plants	Mechanical support
	Amylose	Plants	Energy reservoir
	Dextran	Synthesized by bacteria	Matrix for growth of organism
	Chitin	Insects, crustaceans	Provides shape and form
	Glycosaminoglycans	Connective tissues	Contributes to mechanical support
Polynucleotides	Deoxyribonucleic acid (DNA)	Cell nucleus	Direct protein biosynthesis
	Ribonucleic acid (RNA)	Cell nucleus	Direct protein biosynthesis

Table 1. Incidence and physiological functions of certain natural polymers [12].

DNA is a natural polymer that has a great importance in all living creatures. DNA can be involved in the nucleus of every human cell and determines all of the physical characteristics through genes. Genes consist of a sequence of nucleotides that a specific protein is to be made. Hence the proteins carry out all of the functions of living organisms. Nucleotides are the monomers of DNA and each nucleotide consists of a 5-carbon sugar, a base and a phosphate. There are four bases named as Adenine(A), Thymine(T), Cytosine(C) and Guanine(G). The purines (Adenine and Guanine) have 5 carbons and the pyrimidines (Thymine and Cytosine) have 3 carbons. Purines are nitrogen containing bases consisting of two rings and pyrimidines are nitrogen containing bases with just one ring consisting of carbon and nitrogen. The nucleotides are linked covalently by carbon atoms and covalent bonding occurs between the sugar of one nucleotide and the phosphate in the backbone of the next. Each nucleotide is then paired up with their corresponding base (A to T and C to G). A weak hydrogen bond only holds the two base pairs together which makes DNA very easy to split and replicate. The linked nucleotides become a polynucleotide and named as the polymer of DNA (Figure 2a and 2b) [13].

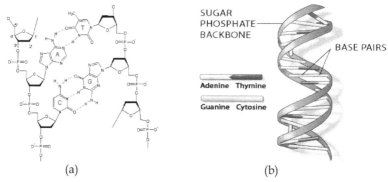

(a)   (b)

Fig. 2. (a) DNA structure [14] and (b) double helix [15].

Polysaccharide is a macromolecule consisting of a large number of monosaccharide chains joined to each other by glycosidic linkages. Polysaccharides are made up chains of monosaccharide molecules which are linked together by glycosidic bonds and occurred by the condensation reaction. The linkage of monosaccharides into chains creates of greatly varying length ranging from chains of just two monosaccharide that make a disaccharide to the polysaccharides which consists of many thousands of the sugars. The polysaccharides play different and significant roles within the biological processes. [16]. The assembly of the polysaccharide is conducted by enzyme-driven reactions. It is the most spacious polysaccharide in nature and acts as a storage carbohydrate in many different parts of a plant. Starch, cellulose, glycogen and chitin are the basic polysaccharides used for medical applications. Starch is a combination of branched and linear polymers of D-glucose molecules generated by plants (Figure 3a). Starch contains only a single type of carbohydrate (glucose) [11].

Cellulose is the most common carbohydrate in the world and is the main substance that forms most of a plant's cell walls. Its structure consists of long polymer chains of glucose units bonded by a beta acetal group. Cellulose structure has mostly a linear chain due to the bond angles in the beta acetal group. The repeating monomer unit in starch structures is alpha glucose. Starch-amylose forms a spiral structure because of the bond angles in the alpha acetal groups (Figure 3b) [17].

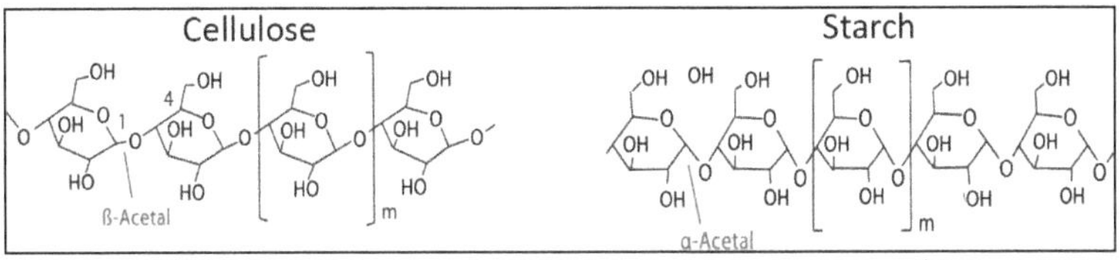

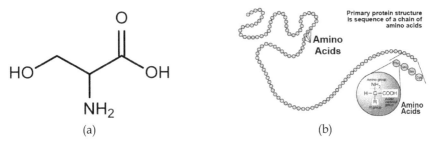

Fig. 3. Structure of (a) starch and (b) cellulose [17].

Collagen is a main structural protein located in the body. It provides homogen strength to the tissues and organs of the body and comprises 90% of the organic matrix of bone together with the mineral, manages through the biomechanical properties and functional integrity of tissues. Collagen that is in the shape of elongated fibrils is mostly included in fibrous tissues such as tendon, ligament and skin, and is also abundant in cornea, cartilage, bone, blood vessels and intervertebral disc. The collagen molecule has a distinctive feature that consists of three polypeptide chains. Proline, hydroxyproline and Gly-Prol-Hyp amino acids are the most common triplet found in collagen. The repeating sequence of these amino acids allows the chains to form a triple-helical structure. The arrangement of triple helices in fibrils is provide high tensile strength to this biopolymer [18].

Proteins are complex biopolymeric structures that are composed of up to 20 different amino acids. These amino acid units are sequenced by the template specific reamer of the polymerization process. Protein chains can contain a few hundreds or thousands of amino acid units (Figure 4).

Fig. 4. (a) Amino acid molecule (serin) and (b protein structure [13].

Chitin is a biopolymer of b(1/4)-linked N-acetyl-D-glucosamine residues (Figure 5a). Chitin is the second most abundant polysaccharide in nature and it is the basic structural component of the exoskeleton of invertebrates such as crustaceans, insects and spiders. It can also be found in the cell walls of most fungi and many algae. Chitosan is obtained from the alkaline deacetylation reaction of chitin and is a linear polysaccharide consisting of N-glucosamine and N-acetyl glucosamine units linked by b(1/4) glycosidic bonds (Figure 5b). The deacetylation degree (DD: glucosamine/N-acetyl glucosamine ratio) usually can vary from 30 to 95% depending on the source.

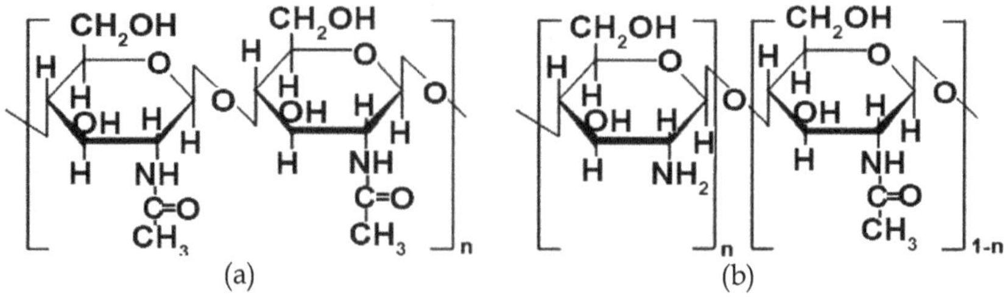

Fig. 5. Chemical structure of chitin (a) and chitosan (b).

Alginate is the name given to linear polysaccharide family found in brown algae and is composed of guluronic and manuronic units. Alginate contains homopolymeric sequences such as polymer of (1-4) - β -D-mannuronopyranosyl and (1-4)-L-guluronopyranosyl units in a copolymer [18]. Table 2 shows some applications of natural biomaterials.

Application	Natural Biomaterial
Artificial heart valves	Bovine pericardium, intact porcine aortic valves
Hernia repair devices	Porcine small intestinal submucoa, porcine urinary bladder mucosa, porcine dermal grafts
Sutures	Catgut (porcine or bovine intestinal wall), porcine dermal grafts
Skin repair/ wound care	Dermal allograft, porcine small intestinal submucoa, porcine dermal grafts
Vascular prostheses	Bovine ureter, porcine small intestinal submucoa, ovine arteries
Urethral repair	Porcine bladder
Breast reconstruction	Dermal allograft
Ligament repair	Dermal allograft, porcine small intestinal submucoa, fetal bovine skin
Spinal fusion/bone healing	Bone allograft

Table 2. Some applications of natural biomaterials [19].

Natural products have many applications in the field of medicine, pharmaceuticals, diagnosis and treatment, food industry and agrochemicals etc. and they have a beneficial for the treatment of particular diseases on the human body. The recent research in natural biomaterials tends to many applications in various biomedical fields.

2.2 Metals

Metallic implant materials have significant economic and clinic importance on the medical applications for a long time. Many of metal and metal alloys such as stainless steel (316L), titanium and alloys (Cp-Ti, Ti6Al4V), cobalt-chromium alloys (Co-Cr), pure metals, precious metals, aluminium alloys, zirconium-niobium and tungsten heavy alloys were used for medical requirements (Table 3). The rapid growth and development of the many specialties of medicine has created whole new medical industry which was to supply more than a trillion dollars of medical products such as dental implants, craniofacial plates and screws;

parts of artificial hearts, pacemakers, clips, valves, balloon catheters, medical devices and equipments, bone fixation devices, dental materials, medical radiation shielding products, prosthetic and orthodontic devices, tools of machining metallic biomaterials. The main criteria in selection of metal-based materials for biomedical applications are their excellent biocompatibility, convenient mechanical properties, good corrosion resistance and low cost [4].

Medical Alloys	Product Forms	Medical Applications
Stainless Steel	Tube, pipe, wire, bar, strip, sheet, plate, screw, profile, clips, fixation devices, nails and pins, joints	Dental and surgical instruments, medical devices
Titanium	Wire, spring	Surgical implants, medical prostheses, dental implants, maxillofacial and craniofacial treatments, cardiovascular devices, surgical instruments
Cobalt-Chromium	Foil, strip, sheet, bracket, screws, surgical instruments	Medical equipment
NiTiNol	Wire, Bar, rod, strip	Surgical instruments, orthodontics, robotics
Pure Metals	Plate, profile	Bone fracture fixation devices
Precious Metals	Forgings, castings	Dental materials
Aluminium Alloys	Fasteners	Medical radiation shielding products
Zirconium-Niobium	Machined parts	Prosthetic and orthodontic devices
Tungsten Heavy Alloys	Prosthetic components	Tools of machining metallic biomaterials

Table 3. Medical metals and alloys for surgical implants and medical devices [5].

At the present time, the compatibility of improved diagnostic instruments and developments about the information on materials are supposed greater significance as well as on medical procedures [8]. Metals are used as biomaterials due to their excellent electrical, thermal and mechanical properties. Metals have some independent electrons which can transfer an electric charge and thermal energy fastly. These mobile free electrons act as the bonding force to hold the positive metal ions together. This strong attraction has proved by the closely packed atomic arrangement resulting in high specific gravity and high melting points of most metals. Since the metallic bond is not essentially directional, the position of the metal ions can be altered without destroying the crystal structure resulting in a plastically deformable solid [9].

Metallic biomaterials such as stainless steel, cobalt-based alloys, titanium and its alloys etc. form either face-centered cubic, hexagonal close-packed or body-centered cubic unit cells at body temperature with ideal crystal lattice structures as shown in Table 4. The most of metal crystals, in contrast to these ideal atom arrangements, contains lattice defects such as vacancies, dislocations, grain boundaries etc. (Figure 6). The presence of point, line and planary defects in metal internal structure has a strong effect on mechanical, physical and chemical properties.

Lattice Systems	Bravais Lattices				Examples
Triclinic	P $\alpha, \beta, \gamma \neq 90°$				$K_2Cr_2O_7$ $CuSO_4.5H_2O$ H_3BO_3
Monoclinic	$\alpha \neq 90°$ $\beta, \gamma = 90°$	C $\alpha \neq 90°$ $\beta, \gamma = 90°$			Monoclinic Sulphur $Na_2SO_4.10H_2O$
Orthorhombic	P $a \neq b \neq c$	C $a \neq b \neq c$	I $a \neq b \neq c$	F $a \neq b \neq c$	Rhombic Sulphur KNO_3 $BaSO_4$
Tetragonal	P $a \neq c$	I $a \neq c$			White Tin SnO_2 TiO_2 $CaSO_4$
Rhombohedral	P $\alpha = \beta = \gamma \neq 90°$				$CaCO_3$ HgS
Hexagonal	P				Graphite (C) ZnO CdS
Cubic	P (pcc)	I (bcc)	F (fcc)		$NaCl$ Zinc Blende (ZnS) Cu

P: Primitive centered that includes the lattice points on the cell corners only; **I:** Body centered which finds one additional lattice point at the center of the cell; **F:** Face centered that consists one additional lattice point at center of each of the faces of the cell; **A, B or C:** Base centered which has one additional lattice point at the center of each of one pair of the cell faces.

Table 4. The distribution of the 14 Bravais lattice types into 7 lattice systems [20].

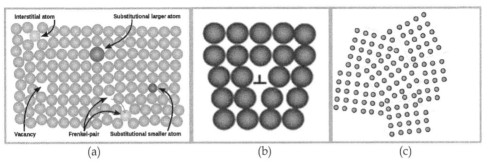

(a)　　　　　　　　　　　　(b)　　　　　　　　　　　　(c)

Fig. 6. Defects of crystal structures: (a) point defects (substitutional or interstitial elements, vacancies) [21], (b) line defects (edge dislocation) [22] and (c) planar defects (grain boundaries) [23].

Interatomic bonding in solids occurs by strong primary ionic, covalent, and/or metallic bonds and weaker secondary interatomic bonding by van der Waals and hydrogen bonds. Metals are characterized by metallic interatomic bonding with valance shell electrons forming an electron cloud around the atoms/ions. As a result of the close positioning of neighboring atoms and the shared valence electrons, the non-directional interatomic bonding and electron movement within metal crystal lattices is easier than in ionic or covalently bonded materials. This fundamental distinctive feature of metals results in the relative ease of plastic deformation as well as the high electrical and thermal conductivities of metals. Most of the metals used for implant production have either close-packed atomic structures with face-centered cubic (fcc) or hexagonal close-packed (hcp) unit cells or nearly close-packed structures forming body-centered cubic (bcc) structures. The equilibrium distance between atoms defining the unit cells of these crystals and the strength of their interatomic bond is determined by intrinsic factors such as atom size and valency as well as extrinsic factors such as temperature and pressure. In addition to ease of deformation to desired shapes, the ability of deforming plastically at increasing loads results in another very important characteristic namely an ability to decrease sharp discontinuities through plastic deformation there by reducing local stress concentrations resulting in relatively high fracture toughness. These desirable features are dependent on proper selection of processing conditions for material and part formation [24].

Type of metal used in biomedical applications depends on functions of the implant. 316L type stainless steel (316L SS) is still the most used alloy in all implants part ranging from cardiovascular to otorhinology. However, when the implant requires high wear resistance such as artificial joints, Co-Cr-Mo alloys is better served. Table 5 summarizes surgical implant alloy compositions used for different biomedical applications [4].

The chemical properties of materials also are due to their atomic bonding types. Atomic bonding type changes the chemical, physical and mechanical properties of metallic materials. The surfaces of metallic medical products can be degraded in living systems and their changed surfaces can release some by-products to biologic medium. As a result of this releasing process, interactions between cell and tissues with metallic implant surfaces occur. For this reason, recent research gives great importance to surface properties of metallic products for the development of biocompatible materials.

Element	316L Stainless Steel (ASTM F138,139)	Co-Cr-Mo (ASTM F799)	Grade 4 Ti (ASTM F67)	Ti-6Al-4V (ASTM F136)
Al	-	-	-	5.5-6.5
C	0.03 max.	0.35 max.	0.010 max.	0.08 max.
Co	-	Balance	-	-
Cr	17.0	26.0-30.0	-	-
Fe	Balance	0.75 max.	0.30-0.50	0.25 max.
H	-	-	0.0125-0.015	0.0125 max.
Mo	2.00	5.0-7.0	-	-
Mn	2.00 max.	1.0 max.	-	-
N	-	0.25 max.	0.03-0.05	0.05 max.
Ni	10.00	1.0 max.	-	-
O	-	-	0.18-0.40	0.13 max.
P	0.03 max.	-	-	-
S	0.03 max.	-	-	-
Si	0.75 max.	1.0 max.	-	-
Ti	-	-	Balance	Balance
V	-	-	-	3.5-4.5
W	-	-	-	-

Table 5. Surgical implant alloy compositions (wt %) [6,7].

The mechanical properties of materials have a great importance during the design of the load-bearing dental and orthopaedic implants. With a few exceptions, the high tensile strength and fatigue limit of metallic materials make them the materials of chosen materials for implants that carry mechanical loads compared with ceramics and polymeric materials. It should be noted that, in contrast to the nanophase, composite nature of their constituent macromolecules. The properties of metallic biomaterials as hard tissue implants are compared below [25]:

Stainless steel (SS)

- cheaper than other metals,
- has high strength, ductility and toughness,
- easy machining,
- can be corrosion problem,
- can be less bone bonding than other metals,
- may create nickel ion sensitivity.

Co-Cr-Mo alloys

- cast alloy forms have lower cost than wrought or forged forms,
- cast alloys forms have higher elastic modulus than wrought or forged forms,
- wrought or forged forms has the highest strength/wear resistance,
- hardest to fabricate,
- may produce cobalt or chromium ion sensitivity/toxicity.

Titanium alloy (Ti-6Al-4V) versus Titanium (Ti) metal

- titanium alloy is stronger than titanium metal,
- both have relatively low elastic modulus,

- neither is as wear resistant as SS or Co-Cr-Mo alloys,
- both have the best corrosion resistance,
- both have excellent bone bonding.

Types of Alloys Used in Medical Devices

Several alloys are used in many medical devices; (1) silver-tin-mercury-copper alloys for dental amalgam fillers; (2) cobalt-chromium alloys for dental applications, cardiac valves, bone fracture and joint components; (3) titanium alloys for conductive leads, screws, joint components and nails and (4) stainless steels for bone fracture and joint components, stents and prosthetic materials (Table 6 and Figure 7).

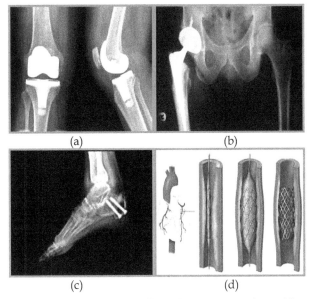

(a)   (b)

(c)   (d)

Fig. 7. (a) Nanostructured diamond coatings [26]; (b) ceramic and metal hip implant [27]; (c) metal ankle implants[28]; (d) coronary artery, metallic stent and stent procedure [29].

Application	Metal
Conductive leads	Titanium and its alloys
Dental amalgams	Silver-tin-copper alloys, gold and platinum fillings
Dental applications	Cobalt-chromium alloys
Fracture plates	Stainless steel, cobalt-chromium alloys
Guide wires	Stainless steels
Heart valves	Cobalt-chromium alloys
Joint components	Cobalt-chromium alloys, titanium alloys
Nails	Cobalt-chromium alloys, titanium alloys
Pacemaker cases	Titanium alloys
Screws	Cobalt-chromium alloys, titanium alloys
Vascular stents	Stainless steel, nitinol

Table 6. Medical applications of metals [30].

Mechanical properties of metals used in medical applications are given in Table 7. Stainless steel has ultimate tensile strength values that range from 480 to 860 MPa and ultimate tensile strain values from 12 to 45%. In comparison, cobalt-based alloys have ultimate tensile strength values that can exceed 1,500 MPa and ultimate tensile strain values between 8% and 50%. Titanium alloys have ultimate tensile strength values up to about 900 MPa and ultimate tensile strain values between 10 and 24%. In comparison to polymers, metals have higher ultimate tensile strength and elastic modulus but lower strains at failure. However, in comparison to ceramics, metals have lower strengths and elastic modulus with higher strains to failure [30].

Type and Condition	Metal	UTS (MPa)	Yield at 2%a	US (%)
F55, F138 (annealed)	Stainless steel	480-515	170-205	40
F55, F138 (cold worked)	Stainless steel	655-860	310-690	12-28
F745 (annealed)	Stainless steel	480 min.	205 min.	30 min.
F75 (cast)	Cobalt-based alloy	655	450	8
F90 (annealed)	Cobalt-based alloy	896	379	30-45 min.
F562 (solution annealed)	Cobalt-based alloy	793-1,000	241-448	50
F562(cold worked)	Cobalt-based alloy	1,793 min.	1,586 min.	8
F563 (annealed)	Cobalt-based alloy	600	276	50
F563 (cold worked and aged)	Cobalt-based alloy	1,000-1,586	827-1,310	12-18
F67	Titanium alloy	240-550	170-485	15-24 min.
F136	Titanium alloy	860-896	795-827	10 min.

Note: UTS = ultimate tensile strength; US = ultimate strain; min. =minimum; a Yield strength at 2% offset in MPs.

Table 7. Mechanical properties of metals used in medical applications [30].

Nowadays, some metal implants have been replaced by ceramics and polymers due to their excellent biocompatibility and biofunctionality. However, the properties of high strength, toughness and durability are required for the metals. On the other side, clinical application of the promising research in using bioactive polymers and ceramics in regenerative medicine is still far away from practice. The future trend seems to combine the mechanically superior metals and the excellent biocompatibility and biofunctionality of ceramics and polymers to obtain the most desirable clinical performance of the implants.

2.3 Polymers

Polymers have been used widely in medicine and biotechnology [31], surgical devices [32], implants [33], drug delivery systems, carriers of immobilized enzymes and cells [34], biosensors, bioadhesives, ocular devices [35], dental materials [36], surface modification [37], biosensors [38], components of diagnostic assays [39], tissue adhesives [40] and materials for orthopaedic and tissue engineering applications [41]. This versatility requires the production of polymers prepared in different structures and compositions and appropriate physicochemical, interfacial and biomimetic properties to meet specific applications. The main advantages of the polymeric biomaterials compared to metal or ceramic materials are ease of manufacturability to produce various shapes such as membranes, films, fibres, gels, sheets, hydrogels, capsules, spheres, particles and 3D-structures (scaffolds) (Figure 8) and ease of secondary processability, reasonable cost and availability with desired mechanical and physical properties. The required properties of polymeric biomaterials are biocompatibility, sterilizability, convenient mechanical and physical properties, and manufacturability as given in Table 8. Different polymeric materials and their biomedical applications are also presented in Table 9. The main types of polymers for biomedical applications derived from natural or synthetic organic sources.

Table 10 presents the most important natural and synthetic origin polymers and their main biomedical applications. The principal disadvantages of these polymers are related to the difficulties in the development of reproducible production methods, because their complex structure often renders modification and difficult purification. In the case of synthetic polymers, these are available in a wide variety of compositions with modified properties.

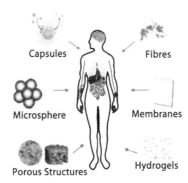

Fig. 8. Natural polymers used in biomedical applications such as tissue regeneration and drug delivery systems [42].

The main disadvantage of synthetic polymers is the general lack of biocompatibility in the majority of cases, often associated with inflammatory reactions [42]. For this reason, the recent researches have focused on the usage possibilities of natural origin polymers such as chitosan, carrageenan, alginate etc.

Property	Description
Biocompatibility	Noncarcinogenesis, nonpyrogenicity, nontoxicity, nonallergic response
Sterilizability	Autoclave, dry heating, ethylenoxide gas, radiation
Physical property	Strength, elasticity, durability
Manufacturability	Machining, molding, extruding, fiber forming

Table 8. Required properties of polymeric biomaterials [9].

Polymer	Application
Poly(methyl methacrylate)	Intraocular lens, bone cement, dentures
Poly(ethylene terephthalate)	Vascular graft
Poly(dimethylsiloxane)	Breast prostheses
Poly(tetrafluoroethylene)	Vascular graft, facial prostheses
Polyethylene	Hip joint replacement
Polyurethane	Facial prostheses, blood/device interfaces

Table 9. Biomedical applications of different polymeric materials [43].

Synthetic Polymer	Main Applications and Comments
Poly(lactic acid) poly(glycolic acid) and their copolymers	Sutures, drug delivery systems and in tissue engineering, biodegradable, regulate degradation time during copolymerization
Poly(hydroxyl butyrate) poly(caprolactone) and copolymers poly(alkylene succinate) etc.	Biodegradable, used as a matrix for drug delivery systems, cell microencapsulation, to change the properties of materials by chemical modification, copolymerization and blending
Polyamides (nylons)	Sutures, dressing, haemofiltration and blending
Polyethylene(low density)	Sutures, catheters, membranes, surgical treatments
Poly(vinyl alcohol)	Gels and blended membranes for drug delivery and cell immunoisolation
Poly(ethylene oxide)	Highly biocompatible, different polymer derivatives and copolymers for biomedical applications
Poly(hydroxyethyl methacrylate)	Hydrogels as soft contact lenses, for drug delivery, for skin coatings and immunoisolation membranes
Poly(methyl methacrylate)	This and its copolymers in dental implants and bone replacements
Poly(tetrafluroethylene) (Teflon)	Vascular grafts, clips and sutures, coatings
Polydimethylsiloxanes	A silicone, implants in plastic surgery, orthopaedics, blood bags and pacemakers
Poly(ortho esters)	Surface eroding polymers, application in sustained drug delivery, ophthalmology
Polyanhydrides	Biodegradable, useful in tissue engineering and for the release of the bioactive molecules

Table 10. Main properties and applications of synthetic polymeric biomaterials [42].

Polymers are large organic macromolecules consist of repeating units called "mers" which are covalently bonded chains of atoms. These macromolecules interact with one another by weak secondary bonds such as hydrogen and van der Waals bonds to form entanglement structure. Polymers exhibit weak thermal and electric properties because of the covalent interatomic bonding within the molecules. The mechanical and thermal behavior of polymers is influenced by several factors, including the composition of the backbone, chemical side groups, chain structures and different molecular weight. Plastic deformation that occurs when applied the mechanical forces cause the movement of macromolecule chains to one another. Changes in polymer composition or structure increase resistance to relative chain movement, so this resistance increases the strength and decreases the plasticity of the material (Figure 9). Substitutions into the backbone that increase its rigidity limit the chain movement. Large side groups also make disentanglement more difficult. Growing macromolecule chain also makes it less mobile and hinders its relative movement [6].

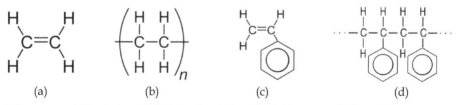

(a)   (b)   (c)   (d)

Fig. 9. Structure of (a) ethylene [44], (b) polyethylene [44], (c) styrene [45] and (d) polystyrene [44].

Polymers can be classified according to the polymerization techniques of the monomers. In bulk polymerization, only the monomer and possibly catalyst and initiator are fed into the reactor without solvent. At the end of the polymerization, process a nearly solid mass is removed as the polymer product. Bulk polymerization is employed widely in the manufacture of condensation polymers, where reactions are only mildly exothermic and viscosity is mostly low thus enhancing ready mixing, heat transfer and bubble elimination. Solution polymerization involves polymerization of a monomer in a solvent in which both the monomer (reactant) and polymer (product) are soluble. Suspension polymerization refers to polymerization in an aqueous medium with the monomer as the dispersed phase. Consequently, the polymer producing from such a system forms a solid dispersed phase. Emulsion polymerization is similar to suspension polymerization but the initiator is located in the aqueous phase (continuous phase) in contrast to the monomer (dispersed phase) in suspension polymerization. Besides, in emulsion polymerization the resulting polymer particles are considerably very smaller than those in suspension polymerization [46].

All polymers follow a degradation sequence in which the polymer is first converted to its monomers and then mineralization of monomers occurs. Most polymer molecules have too large size to pass through the cellular membranes, so firstly they have to depolymerize to smaller monomers before microbial cells absorbe them in biological system. The begining of polymer degradation during the chemical hydrolysis process can result different physical, chemical and biological forces. Physical forces such as heating/cooling, freezing/melting or wetting/drying can cause mechanical cracking of polymeric materials. These physical forces deteriorate the polymer surfaces and reveals the new surfaces for reaction with chemical and biochemical agents. This feature is a critical phenomenon in the degradation of solid polymers but the chemical and biological forces of fluid polymers are more important [47,48].

In recent years, advances of innovative technologies in tissue engineering, regenerative medicine, gene therapy and drug delivery systems have promoted through the need of new biodegradable biomaterials. Biologically and synthetically derived biodegradable biopolymers have attracted considerable attention. Polysaccharides and proteins are typical biologically derived biopolymers while aliphatic polyesters and polyphosphoesters are typical synthetic biopolymers. Specific biopolymers needed for in vivo applications are required because of the diversity and complexity of in vivo environments. Nowadays, synthetic biopolymers have become attractive alternative materials for biomedical applications for three reasons for most biologically derived biodegradable polymers: (1) may induce an immune response in the human body; (2) difficulty in chemical modifications; (3) change of the bulk properties after chemical modifications. To achieve the specific properties, properly designed synthetic biopolymers require further modifications without altering the bulk properties [49]. Table 11 lists the potential applications of these biodegradable polymers in tissue engineering.

Polymer	Tissue engineering
Polyanhydrides	Bone
Polyurethane	Vascular, Bone
Polyelectroactive materials	Nerve
Polyphosphoester	Bone
Poly(propylene fumarate)	Bone
Polyesterurethane	Genitourinary

Table 11. The potential applications of biodegradable synthetic biopolymers [49].

result in long-distance and three-dimensional crystalline structures but glass materials do not exhibit long-distance order. The electrons in ionic and covalent bonds are circumscribed between the constituent ions/atoms of the metallic bonds; therefore ceramics show the nonconductive property. The strong ionic and covalent bonds make ceramics hard and brittle, because the planes of atoms/ions cannot move one through another. For this reason, ceramics and glasses are sensitive to the presence of cracks or other defects during plastic deformation. The ionic and/or covalent nature of ceramics also influences their chemical behavior [6].

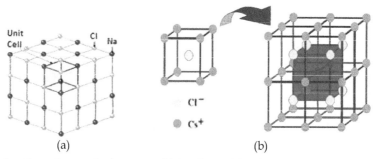

(a)   (b)

Fig. 11. Unit cells and crystal structures of (a) sodium chloride (NaCl) [55] and (b) cesium chloride (CsCl) [56] crystals.

Although ceramics and glasses resist to corrosive effects when exposed to the physiological environment, components of the ceramic structure make them sensitive to degradation media. The rate and mechanism of degradation depend on the structural properties of ceramic. Although alumina ceramic has mostly a bioinert character, its strength decreases in time during the immersion in simulated body fluids and after implantation. Some cracks and crack propagation may accelerate dissolution of impurities so ceramic material degrades in the biological structure.

Bioactive ceramics and glasses are also degraded in the body. [52]. Generally, degradation rate of materials depends on material composition, their functions and components of biological medium. The biodegradable (resorbable) ceramics are used for applications such as drug delivery systems, repairing of damaged or diseased bone, bone loss, filling the implant system vacations, donor bone, excised tumors, repairing vertebra, herniated disc surgery, treatment of maxillofacial and dental defects [53].

The brittleness and poor tensile strength properties of ceramic and glass implants are the main disadvantages. Although they can have exceptional strength during the compression loadings, ceramics and glasses fail at low stress under the mechanical loading such as tension or bending. Alumina and zirconia have the highest mechanical properties among biomedical ceramics. Low friction coefficient and wear rate are other advantageous properties of alumina. These properties of alumina and zirconia ceramics are used as a load-bearing surface in joint replacements [6].

Most of scientific works have been devoted to the interfacial reactions of biological systems with hydroxyapatite ceramic having chemical structure very similar to the mineral phase composition of bone. Hydroxyapatite is a popular surface coating material for stainless steels, titanium and its alloys implants. In the recent time, researchers have investigated the probability of its use in composite forms that integrate polymers with ceramic or metal/ceramic combinations. There are significant researches performed on coating techniques, in-situ and biomimetic synthesis of apatites and the implications for ceramic properties and microstructure. The thermal and chemical stability of ceramics, high strength, wear resistance and durability properties contribute to making ceramics good candidate materials for surgical implants [57].

Calcium phosphate-based biomaterials and bioceramics are now used in a number of different applications throughout the body, covering all areas of the skeleton. Applications include dental implants, transdermic devices and use in periodontal treatment, treatment of bone defects, fracture treatment, total joint replacement, orthopaedics, cranio-maxillofacial reconstruction, otolaryngology and spinal surgery depending upon a bioresorbable or a bioactive material (different calcium orthophosphates). Figure 12 shows some commercially available ceramic medical products [58] and Table 14 also presents calcium phosphate based ceramics arranged by Ca:P ratio [54].

(a)   (b)   (c)   (d)

Fig. 12. Ceramic products for medical applications (a) ceramic crown [59]; (b) hydroxyapatite block ceramic [60]; (c) ceramic implant systems [61]; (d) bio-eye hydroxyapatite orbital implants [62].

The mechanical properties of calcium phosphates and bioactive glasses make them unsuitable as load-bearing implants. Clinically, hydroxyapatite has been used as filler for bone defects and as an implant in load-free anatomic sites such as nasal septal bone and middle ear. In addition to these applications, hydroxyapatite has been used as a coating material on metallic orthopaedic and dental implants to promote their fixation in bone. In this case, the fundamental metal surfaces to the surrounding bone strongly bonds to hydroxyapatite. Delamination of the ceramic layer from the metal surface can cause serious problems and results in the implant failure [6]. Intracorporeal implantable bioceramics should be nontoxic, noncarcinogenic, nonallergic, noninflammatory, biocompatible and biofunctional in the biological structure.

Chemically, calcium orthophosphate bioceramics are based on HA (hydroxyapatite), β-TCP (β-tricalcium phosphate), α-TCP (α-tricalcium phosphate) and/or BCP (biphasic calcium phosphate). The BCP concept is determined by the optimal phase ratio between HA with α-TCP or β-TCP phases. Desirable bone tissue should have two clinical requirements: (1) pores size of the bone tissue grafts should be about 100 μm for biodegradation rate comparable to the formation of bone tissue between a few months with about two years and (2) their mechanical stability should be sufficient for clinical demand. HA is a more stable phase under the physiological conditions compared to α-TCP and β-TCP phases, as it has a lower solubility and a slower resorption rate. Calcined HA-based implants are applied for

Name	Abbreviation	Formula	Ca/P
Tetracalcium phosphate	TetCP	$Ca_4O(PO_4)_2$	2.0
Hydroxyapatite	HAp	$Ca_{10}(PO_4)_6(OH)_2$	1.67
Amorphous calcium phosphate	ACP	$Ca_{10-x}H_{2x}(PO_4)_6(OH)_2$	
Tricalcium phosphate (α, β, γ)	TCP	$Ca_3(PO_4)_2$	1.50
Octacalcium phosphate	OCP	$Ca_8H_2(PO_4)_6.5H_2O$	1.33
Dicalcium phosphate dihydrate (brushite)	DCPD	$CaHPO_4.2H_2O$	1.0
Dicalcium phosphate (monetite)	DCP	$CaHPO_4$	1.0
Calcium phosphate (α, β, γ)	CPP	$Ca_2P_2O_7$	1.0
Calcium pyrophosphate dihydrate	CPPD	$Ca_2P_2O_7.2H_2O$	1.0
Heptacalcium phosphate	HCP	$Ca_7(P_5O_{16})_2$	0.7
Tetracalcium phosphate diacid	TDHP	$Ca_4H_2P_6O_{20}$	0.67
Calcium phosphate monohydrate	MCPM	$Ca(H_2PO_4)_2.H_2O$	0.5
Calcium metaphosphate (α, β, γ)	CMP	$Ca(PO_3)_2$	0.5

Table 14. Calcium phosphates arranged by Ca:P ratio [54].

repair of the bone defects after implantation for many years, bioceramics made of β-TCP, α-TCP, CDHA (calcium-deficient hydroxyapatite) or BCP are more preferable for medical purposes. According to both observed and measured bone formation parameters, calcium orthophosphates were arranged as low sintering temperature BCP (rough and smooth) ≈ medium sintering temperature BCP ≈ TCP > calcined low sintering temperature HA > non-calcined low sintering temperature HA > high sintering temperature BCP (rough and smooth) > high sintering temperature HA (calcined and non-calcined). Figure 13 shows some randomly chosen examples of commercially available calcium orthophosphate bioceramics for use as bone grafts [63].

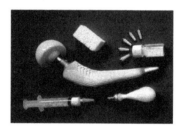

Fig. 13. Various commercial calcium phosphate-based bioceramic products [63].

Bioresorption and bioactivity are the most important properties of calcium phosphate-based biomaterials. Even though the bone mineral crystals have a very large surface area, beside this, calcium phosphate bioceramics present lower surface area and strong crystal bonding. Because of this reason, disintegration of particles into crystals and the dissolution of crystals are involved in a resorption process. While cell activity forms the dissolution stage of crystals, insoluble calcium phosphates such as apatite or calcium pyrophosphates cannot be eliminated easily by cells and can reveal infected properties in other tissues. High sintering temperature (>900°C) decreases the resorption rate of HA ceramics, on the contrary (<900°C) increases [52]. Although HA sintered ceramics show osteoconductivity, their bioresorbability is low and HA remains in the body for a long time after implantation. Implanted materials should exhibit resorbable property through the bone regeneration followed by complete substitution for the natural bone tissue after stimulation of bone formation. Therefore, recent research has been tend to TCP ceramics as scaffold materials for bone regeneration [64].

Calcium phosphate ceramics exhibit nontoxic behavior to tissues, bioresorption and osteoinductive property. Since ceramic/ceramic, ceramic/metal, ceramic/polymer composites include the different solid particle stiffeners, they need to the choice of more materials for implant applications. Consequently, the biological activity of bioceramics has to be known by various in vitro and in vivo studies and also data on mechanical feature would determined using by standard test methods to make the implant application and the choice of the bioceramic depending on the site of implantation easier [52]. Eventually, implantable bioceramics should be nontoxic, noninflammatory, nonallergic, noncarcinogenic, biocompatible and biofunctional for its working time in the host tissue.

In consequence of bioceramics have generally structural properties and functions; they were used as joint or tissue replacements, coating materials to improve the biocompatibility of metal implants, temporary structures and frameworks to rebuild the damaged tissues in the body, drug delivery system for the treatment of damaged living organism.

2.5 Composites

Composites are engineering materials that contain two or more physical and/or chemical distinct, properly arranged or distributed constituent materials that have different physical properties with an interface separating them. Composite materials have a continous bulk phase called the matrix and one or more discontinuous dispersed phases called the reinforcement which usually has superior mechanical or thermal properties to the matrix. Separately, there is a third phase named as interphase between the matrix and reinforced phases such as coupling agent coated on glass fibers to achieve adhesion of glass particles to the polymer matrix [65].

Solid materials basically can be classified as polymers, metals, ceramics and carbon as a separate class. Both reinforcements and matrix materials are divided into four categories. Composites are usually classified as the type of material used for the matrix. The four main categories of composites are polymer matrix composites (PMCs), metal matrix composites (MMCs), ceramic matrix composites (CMCs) and carbon/carbon composites (CCCs). Recently, PMCs are the most commonly preferred class of composites. There are important medical applications of other types of composites which are indicative of their great potential in biomedical applications [66].

The composite material usually includes reserved distinct phases that are separated on a scale larger than the atomic size. A synthetic composite material also consisted of continuous polymeric matrix phase and a ceramic reinforcement phase like natural biological materials such as bone, dentin, cartilage, skin. The properties of this phase such as the elastic modulus are significantly altered when compared with a homogeneous structured material. Reinforced polymer matrices are fiberglass or natural materials such as bone. Bone, wood, dentin, cartilage and skin tend to be natural biological composite materials, beside these; natural foams include lung, cancellous bone, wood, sponge etc. Natural composites have hierarchical structures particulate, porous and fibrous structural features which are seen on different micro-scales [9].

Composite materials consist of mixtures of polymers, metals and ceramics to form materials such as fiberglass, a mixture of glass fibers coated with a polymeric matrix. In recent years, scientific research tends to develop biomedical composite materials because of these materials present new alternative solutions for load-bearing tissue components. Composite materials are used limitedly in medicine for example high-modulus carbon fibers embedded in a polymeric matrix such as poly(lactic acid) and carbon-fiber-reinforced high molecular weight polyethylene are widely studied for tendon, ligament, joint and facial implants [30].

The structural properties are main feature of composite materials. Composites are different from homogeneous materials because of considerable control can be exerted over the larger scale structure, and hence over the desired properties. The properties of a composite material depend on the shape of the reinforcements, the volume fraction of them and interaction level of the interfaces of constituents. The categories of basic dispersed phase shape in a composite material are (1) particle; (2) fiber; and (3) platelet or lamina as shown in Figure 14. The dispersed phases may vary in size and shape within all category. For example, particulate dispersed phases may be spherical, ellipsoidal, polyhedral or irregular. If any phase consists of voids filled with air or liquid, the material is known as a cellular solid. If the cells have polygonal shapes, the material is a honeycomb form; if the cells have polyhedral shape, material has a foam form. These particulate shapes are necessary to construct the biomaterial to distinguish the above structural cells from biological cells which occur only in living organisms. Moreover, produced composite structure has to include random orientation and preferred orientation [67].

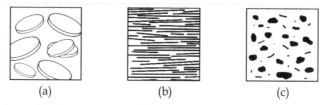

(a)   (b)   (c)

Fig. 14. Morphology of basic composite inclusions: (a) particle, (b) fiber, (c) platelet [67].

Metals have high strength, ductility and wear resistance properties. Most of many metals have low biocompatibility, corrosion, high stiffness and density, and metal ion releasing when compared to tissues. However the ceramic materials are exhibited good biocompatibility and corrosion resistance and high compression resistance, they also show brittleness, low fracture strength and production difficulties. But polymer composite materials provide alternative route to improve many undesirable properties of homogenous materials mentioned above.

In general, tissues are grouped as hard and soft tissues, the bone and tooth are examples of hard tissues and skin, blood vessels, cartilage and ligaments are some of the soft tissue examples. The hard tissues have higher elastic modulus and stronger tensile strength than the soft tissues. Metals or ceramics chosen for hard tissue applications together with polymers preffered for the soft tissue applications have to exhibit structural and mechanical compatibility with tissues. The elastic modulus of metal and ceramic materials is 10-20 times greater than values of the hard tissues. This difference is a major problem in orthopedic surgical materials between the bone and metallic or ceramic implants. Because of load sharing between the bone and implant, it can create different stress distribution directly related to their stiffness [68].

Bone consisting of apatite mineral on collagen fibrils in different mass ratios form a composite material structure (Figure 15). Composite structure with a bioactive component will induce tissue growth to the implant and the formation of a strong bond between the tissue and implant after implantation process. [69].

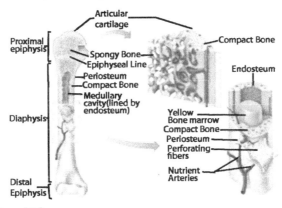

Fig. 15. Bone structure [70].

Shape, size and distribution, volume ratio, bioactivity properties of the reinforcement or matrix phases; matrix properties such as molecular weight, grain size and distribution; ratio of the reinforcement in the matrix and reinforcement-matrix interfacial situation are largely affected to the properties of biomedical composites. Among these factors, properties of constituent materials have significant influence. However, architectural features of the composites such as the reinforcement percentage, distribution and orientation, etc. and reinforcement-matrix bonding condition also have strongly important roles. The mechanical and biological performance of bioactive composites relates to response various clinical requirements. The physical characteristics of the reinforcement such as shape, size and size distribution determines the mechanical properties of a composite. Spherical reinforcement shape which determined by mathematical modelling results achieved in the idealised situation provides good mechanical behavior for particulate composite material (Fig. 16a). Substantially, reinforcing bioactive particles may have to different shapes like irregular, platelet or needlelike. And the irregular shapes facilitate to the penetration of the molten polymer into gaps on the particle surface during high temperature composite processing. Thus, mechanical interlock between reinforced particle and polymer effectively form at the ambient or body temperature and even if a tensile stress is applied (Fig. 16b). The reinforcing bioactive glass or glass–ceramic particles which are made via the conventional glass–making method such as melting and quenching have sharp corners and take up the shape shown in Fig. 16c. These irregular shapes with sharp corners that cause stress concentration in the composites around are not preferred. The platelet shape is rarely encountered for particles in bioactive composites (Fig. 16d). Generally, particles prepared using by the precipitation procedure have to nanometer size and the needlelike shape are directly used for the composite materials (Fig. 16e) [69].

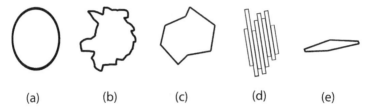

(a)   (b)   (c)   (d)   (e)

Fig. 16. Shapes of bioceramic particles for biomedical composites: (a) spherical, (b) irregular flake, (c) irregular flake with sharp corners, (d) platelet, (e) needlelike [68].

Polymer composites are combination of two or more materials such as fillers, polymer matrix and binders especially structurally supplemental substances as if metal, ceramic, glass and polymer combine to produce structural or functional properties none of any specific component. Although the constituents protect identities of them and do not dissolve or merge completely into one another, they protect the rules of the integrity. Normally, the components of the composite body form an interface between two phases. The composites are divided in four main categories as composites with particles; composites reinforced with fibers; layered composites and sandwich structured composites [71].

The most contribute factors to the engineering performance of the composite are: (1) materials that make up the individual components; (2) quantity, form and arrangement of the components and (3) interaction between the components. The shape, size, orientation, composition, distribution and manner of incorporation of the reinforcement system in a composite material strongly determine the desirable properties of material. Composite materials can also be simply classified as fiber-reinforced and particle-reinforced composites. Thus, there are four classes of composites: polymer-matrix composites (PMCs), ceramic-matrix composites (CMCs), or metal-matrix composites (MMCs). MMCs are prefferred rarely in biomedical applications and mostly used for high-temperature applications [65].

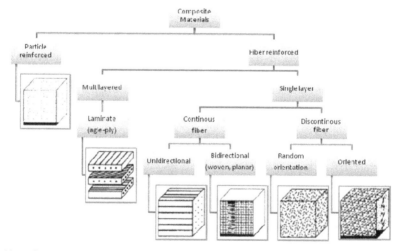

Fig. 17. Classification of composite materials [65].

Figure 17 shows main types of biocomposites according to their reinforcement forms. Three kinds of reinforcements as short fibers, continuous fibers and particulates are typically used in preparation of the medical composites such as screws and total hip replacement stems made from short fiber reinforcements [68].

A typical external fixation system comprises of wires or pins that are pierced through the bone and keep under high tension by screws to the external frame. The wires oriented at different angles across the bone and their tensions are adjusted to provide necessary fixation rigidity. The external fixators are designed with high rigidity and strength to ensure the stability. Traditional designs are made of stainless steel which are heavy and cause patients to feel uncomfortable carrying the system for several months. CF/epoxy composite material is a good alternative for external fixator's construction and gain currency owing to the fact that their lightweight, sufficient strength and stiffness [68]. Biocomposites used in the body are shown in Figure 18.

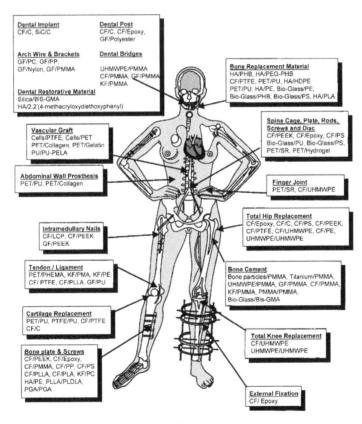

Fig. 18. Biocomposites used in the body [68].

The matrix determines the environmental durability of the composite by resisting chemical, hygroscopic and thermal stresses and protecting the reinforcement from these stresses. The matrix also greatly influences the processing characteristics of a composite material. Common matrices and reinforcements in biomedical composites are listed in Table 15. Thermoplastics are selected as matrix materials due to their nonreactive nature to produce biomedical composite materials. In spite of the thermosets have to inert feature; reactivity, flexibility and strength properties of some thermoplastics have redesigned to provide biodegradable behavior. Resorbable matrices are temporarily convenient to prepare a composite implant, but they are attractive to design a stiff reinforcing material that have a comparable degradation. Good compressive properties and bioactive possibilities of ceramics have also presented new matrix opportunities to produce the modified composite structures [65]. New matrix materials are developed constantly instead of ready for sale materials for medical applications.

Composites can be made from constituents that have similar linear expansion constants. If they have distinct linear expansion constants, contact area (interface) between reinforcement and matrix materials can generates large voids through the contact surface, therefore bone tissue engineers which produce the composite scaffolds have to behave more sense for the selection of the polymeric and bioceramic materials. The nature of the materials in a composite material usually requires the essential mechanical and/or chemical properties for the polymers and ceramics (and glasses). Materials that have the ability to degrade in vivo are ideal candidates for composite scaffolds which gradually degrade. During the degradation of calcium phosphate or bioactive glasses, the polymeric byproducts that form due to low pH values buffered to prevent the formation of inconvenient environment conditions for cells.

Matrix	Fibers	Particles
Thermosets	*Polymers*	*Polymers*
Epoxy	Aromatic polyamides	Polylactide, and its copolymers
Polyacrylates	Poly(ether ketones)	with polyglyocolide
Polymethacrylates	Polyesters (aramids)	Collagen Silk
Polyesters Silicones	UHMWPE	*Inorganic Glass*
Thermoplastics	Polyesters	Alumina
Polyolefins (PP, PE)	Polyolefins	*Organic*
UHMWPE	PTFE	Polyacrylate
Polycarbonate	*Inorganic*	
Polymethacrylate	Carbon	
Polysulfones		Glass
Inorganic	Hydroxyapatite	
Hydroxyapatite	Tricalcium phosphate	
Glass ceramics	*Resorbable polymers*	
Calcium carbonate ceramics		
Calcium phosphate ceramics		
Carbon Steel		
Titanium		
Resorbable polymers		
Polylactide, polyglycolide and their copolymers		
Polydioxanone, Poly (hydroxy butyrate)		
Alginate, Chitosan, Collagen		

Table 15. Constituents of biomedical composites [1].

Bioceramic and glass materials are mechanically stronger than polymers and have a critical significant in providing mechanical stability to construct prior to synthesis of new bone matrix by cells. However, ceramics and glasses tend to destructive failure depending on to their intrinsic brittleness and defect sensitivity. It is required to obtain good chemical and/or physical bonding between polymer and inorganic phase to optimise the biological and mechanical performance of bioactive polymer/ceramic composites.

Table 16 lists selected typical biodegradable and bioactive ceramic/glass-polymer composites that are designed for bone tissue engineering scaffolds and their mechanical properties. Literature reported an important composite scaffolds group that are tailored combinations of bioglass particles and biodegradable polymers such as PLGA, PDLLA, PHB having high application potential. These composites with porous structure have such mechanical properties that are close to cancellous bone and the high bioactivity is conferred by the bioglass particulate filler.

Biocomposite		Percentage of Ceramic (wt%)	Porosity (%)	Pore size (µm)	Compressive (C) Tensile (T) Flexural (F) Strength (MPa)	Modulus (MPa)
Ceramic	Polymer					
Non-crystalline CaP	PLGA	28-75	75	>100		65
ß-TCP	Chitosan-Gelatine	10-70	-	322-355	0.32-0.88 (C)	3.94-10.88
ß-TCP	PLGA	30	-	400(macro) 10(micro)	-	-
HA	PLLA	50	85-96	100x300	0.39 (C)	10-14
	PLGA	60-75	81-91	800-1800	0.07-0.22 (C)	2-2.75
	PLGA		30-40	110-150	-	337-1459
nHA	PA	60	52-70	50-500(macro) 10-50(micro)	13.20-33.90 (C)	0.29-0.85
HA	PCL	25	60-70	450-740		76-84
HA	PLAGA	50-87	-	-	80 (C)	Up to 120
Bio-glass®	PLGA	75	43	89	0.42 (C)	51
	PLLA	20-50	77-80	100(macro) 10(micro)	1.5-3.9 (T)	137-260
	PLGA	0.1-1	-	50-300	-	-
	PDLLA	5-29	94	100(macro) 10-50(micro)	0.07-0.08(F)	0.65-1.2
CaP glass	PDLLA	20-50	93-96.5	80-450	-	0.05-0.2
A/W	PLA-PDLLA	40	93-97			
Phosphate		20-40	85.5-95.2	98-154	0.017-0.020 (C)	0.075-0.12
Glass	PDLLA	-	-	-		
Human cancellous bone	-	-	-	-	4-12 (C)	100-500

Table 16. Typical biodegradable and bioactive ceramic-glass-polymer composites for bone tissue engineering applications and their mechanical properties [73].

Many studies suggest well-dispersed nanostructured composites that may offer surface and/or chemical properties closer to native bone and therefore they might represent ideal substrates to support bone regeneration. Recently, nanosized bioactive ceramic and glass particles have become available which can be considered as ideal fillers for tissue engineering scaffolds. However, problems associated with poor interfacial bonding and particle agglomeration may be more pronounced when using nanosized particles. Coupling agents have been employed as well as titanates and zirconates to improve the bonding between inorganic particles and matrix [72].

Fibre reinforced composite materials are widely prefferred for hard-tissue applications such as skull reconstruction, bone fracture repair, total knee, ankle, dental, hip and other joint replacement applications [73]. Considerably, the mechanical properties of composite materials depend on the direction of loading and the volume fraction of fibers. Low volume fraction of fibers decreases the modulus of the composite and requires a critical volume fraction approach for the modulus of the composite fibers. The modulus of cartilage and collagen fiber in a matrix of glycosaminoglycan and the tensile strength of skin, thick collagen fiber and glycosaminoglycan can be given as examples of the mechanical properties of composites. Biomedical composites which show continuous progression are very attractive and promising engineering materials made from two or more constituents with different physical properties [74].

3. Components of living systems

3.1 Structure and properties of cells and tissues

Cells the smallest living unit of an organism and each cell maintains all necessary functions to survive and represent the organism's genetic code in its linear chemical cord (deoxyribo nucleic acid -DNA). They also evolve by the time or environmental changes; keep those records of experiences and can transfer this information to surrounding neighbor cells or next generations. Cells act like a power plant that requires raw material to produce energy and function. Each cell has a programmed life-span and predetermined function organized by the genetic code. According to their encoded function cells differentiate and develop specific features both functionally and physically during the intrauterine life.

All cells have certain components that enable them to maintain their vital life processes. Those components are responsible to carry out different functions of the cell [75,76]. Cell membrane is a bilayered phospholipid barrier that separates the interior colloidal suspension (cytoplasm) of all cells from the outside environment (Figure 19).

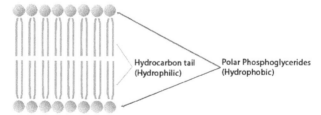

Fig. 19. Lipid bilayer model of a cell membrane. Circular polar portions making internal and external borders are made of phosphoglycerides (hydrophobic lipid), while inner portion of the molecule is hydrocarbon (hydrophilic).

Exterior and interior surfaces of cell membrane are lined with lipid pole of phospholipid molecules, while hydrophilic phospho-poles locate in the middle zone of the membrane. Randomly spread protein molecules, acting like a gated between inner and outer spaces, located in that bilayered cell membrane, of which main function is transportation of some molecules (Figure 20). Carbohydrates take place at cell membrane in combination with proteins as glycoproteins and fat acids as glycolipids. Both these carbohydrate containing molecules (glycocalyx) cover outer surface of cell and repulse negative charged molecules or cells.

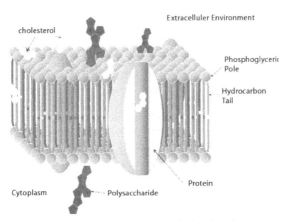

Fig. 20. Fluid-mosaic model of cell membrane. Phospholipid is the major component of the bilayer lipid structure. Protein molecules float within this bilayered structure.

Filtering substances is a selective operation, which essential to maintain homeostatic environment inside the cell to keep vitality of the organism. Even though cell membrane is selective barrier fulfilling passive diffusion, it may not be able to keep the equilibrium between inner structure and outer environment (homeostasis). Certain ions or molecules may require to be moved in or out with enforcement of transporter molecules of cell. This active transportation procedure requires energy.

Common components of functioning cells (organelles) are:

- Nucleus: contains DNA and controls functions of cell,
- Mitochondria: power generator of the cell and control water and mineral level of cytoplasm and responsible from metabolic functions,
- Endoplasmic reticulum: Tubular network among nucleus and cell membrane, responsible from inner transportation and storage,
- Golgi apparatus: Responsible from sorting and storage of protein products of the cell,
- Lysosome: Digestion of proteins, lipids, and carbohydrates and transportation of waste materials to cell membrane,
- Ribosomes: Function for protein synthesis.

Organized gathering of similar cells providing a specific function as a part of organ or organism is called as tissue. Tissues have their own purposeful union to fulfil their main function. Several types of tissues assemble in structural unit to serve a common function and named as "organ". In an organ usually there is a main tissue (parenchyma), which executes the organ function. In order to maintain vital functions and integrity of parenchyma supportive tissue (stroma), vessel tissues and nerve tissues take part in an organ. "Organ system" term is used for collection of several organs working together to fulfil a function, for example digestive system [75,76].

3.2 Biomolecules and their behaviors

Biomolecules are main components of living cells (also their raw digestion materials) and at the same time their end-products, can also have an important role in signal transmission inside the cell or among other cells. Those molecules are also responsible from many functions like regeneration, survival and death. Several classifications can be made for their structures and functions. They can be a single molecule or in polymer form as well.

Roughly saccharides, lipids, amino acids are the basic counterparts of more complicated molecules like proteins, vitamins, hormones and enzymes. Fundamental functions like digestion and replication are performed in a programmed cascade via several molecules under the control of DNA molecule. DNA is a double stranded polymer of the same four types of monomers (nucleotides). A base molecule, which may be either adenine (A), guanine (G), cytosine (C) or thymine (T) binds to a sugar (deoxyribose) with a phosphate group attached to it, forms this nucleotide monomer, which also can be called as building block of a single DNA strand (Figure 21). Those monomers bind one another with sugar phosphate linkages.

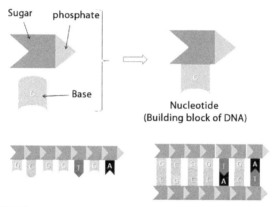

Fig. 21. Structure of DNA.

A DNA strand attached its pair with weak hydrogen links (adenine to thymine, cytosine to guanine), which permits two DNA strands to be separated without breakage. Each those DNA strands are templates for the synthesis of a new DNA strands during cell division to produce new cells.

DNA has more responsibilities than replication but also synthesis of protein and ribonucleic acid (RNA) to maintain its vital functions. This process begins with transcription of some DNA segments and forms RNA polymers that are used as templates for several molecules like proteins or enzymes. In other words DNA contains templates of all molecules that it is responsible to produce.

Amino acids, in other terms backbones of protein polymers, present diversity from DNA or RNA and 20 types of amino acids are used as base monomer for protein (polypeptide) synthesis. This variety brings different structural and behavioral nature diversity for proteins, like maintaining structures via enzymatic activity, movement of cell, sensing any inner or outer signal. According to their genetically encoded nature each protein molecule is responsible from a specific function. Enzymatic action of proteins is mainly dependent their amino acid arrays and location which determines the nature of the enzyme (destruction, digestion, catalization or synthesis function) [76].

3.3 Host response to materials

3.3.1 Inflammation, wound healing, and the foreign body response

Simply, tissue response to any kind of injury can be named as inflammation. Injury is the starting point of protective biomechanism, which triggers wound healing to restore the damaged part of organism.

- Physical trauma, like traffic accidents,
- Thermal injuries, like sun or fire burns and electrical burns,
- Chemical injuries, like contact with acidic or alkali solutions or poisoning or chemically irritant material, and
- Biologic invasion (infections) by microorganisms, like bacteria, virus or fungi can cause tissue injury. Owing that information; it is important to emphasize a common mistake that infection is not a synonym for inflammation, but one cause of inflammation.

Following sudden onset of signal injured vessel system and some local blood cells produce initial response by releasing some biochemical molecules that give first protective reaction to the injury cause and some other produce a chemical signal to attract the white blood cells, which are responsible from body defence, to let them migrate to the injury site. Inflammation stage can also be considered as the first step of wound healing. If the integrity of vessel is interrupted by the injury platelets (thrombocytes) begin to adhere the wound edges and one another to produce clot plug. By the aggregation of platelets they also release some molecules trigger clotting mechanism to produce clot plug. Those vessels or platelet originated molecules increase vessel permeability leading transfusion water from vessel lumen into the injured tissue site, which causes swelling of the tissue (edema). To ease arrival of chemically induced migrating (chemotaxis) of white blood cells; those released biomolecules also loosen muscles surrounding vessels, which are responsible to control blood flow via altering diameter of vessels, and increase vessel diameters. Therefore, the site looks reddish due to dilated blood vessels. Fibroblasts, likewise white blood cells, have the ability of chemotaxis to infiltrate the injury area to start reparatory phase. Some of these released molecules are pain mediators to warn our nerve system to attract our attention to take care our wound.

As the white blood cells arrive the site they begin to struggle with the target caused injury and beside begin to resorb the blood clot to restore vessel lumen, so that blood stream to be maintained. Some white blood cells release molecules to neutralize the threat chemically and some others try to resorb the foreign material in small pieces to relive it (see section 3.4.2). Simultaneously fibroblasts begin to produce connective tissue, which will be replaced with clot until local tissue regeneration takes place. Epithelial cells (responsible to cover our body and make a barrier in between outer environment and inner medium) begin to replicate and grow to create a new cover over the wound area [75,76].

As the source of injury (foreign body) is eliminated and the tissue is regenerated and take the action again wound healing process is completed. However, if the foreign body, sometimes implant or biomaterial, remains inside the body. On that point; it is critically important that whether the material giving threatening chemical signals, or not. If the remaining material is biocompatible, following the initial acute response, it can get along well with the surrounding tissues and may survive inside the organism.

On contrary if the chemical signals of implanted material are recognized as a threat and cannot be terminated, defence system activates rejection mechanism (foreign body reaction) to withdraw the foreign body. Inflammatory reaction continues beside work out of fibroblasts, which knit a connective tissue net around the foreign body to isolate and put this material in quarantine for any possible further threats. As release of inflammatory biomolecules continues swelling, pain and reddish appearance sustains until the withdrawal. This process is called as foreign body reaction and presents a chronic development.

3.3.2 Immunology and the complement system

Our environment is rich of bacteria, viruses, fungi, and parasites and we are in contact with these external threats continuously via our skin, eyes, digestive, respiratory and genitourinary systems. Each attack of those organisms could have been lethal for us, if our body had not got any protection mechanisms. Defence system of human organism against hazardous factors is immune system. This system is not only responsible to fight with those invader organisms, but also has the obligation to neutralize chemical agents sensed as a threat for the organism. This system is sensitive to the dangers developing internally, like tumoral pathologies and sometimes can struggle with our natural structure itself.

Immune system is consist of two major components; cellular and biochemical reactions. Cellular reaction is carried out by the white blood cells produced partly bone marrow and lymph vessels. Polymorphonuclear neutrophils, polymorphonuclear eosinophils, polymorphonuclear basophils, monocytes, lymphocytes, and plasma cells act like soldiers in this battle. They destroy invading microorganisms by engulfing (phagocytosis), especially neutrophils and macrophages.

Second defence mechanism to eliminate or inactivate biologic or chemical agents is biochemical defence. This neutralization process can be innate or adaptive (acquired). Some natural secretions of our organ systems, like acid release of stomach, some digestive enzyme secretion of several organs of digestive system can kill the bacteria entered to our digestive canal. A mucolytic polysaccharide, lysozyme, which is a component of sweat and tear, can dissolute the microorganisms or some basic polypeptides can react with and inactivate certain types of gram-positive bacteria.

The complement system of about 20 proteins including serum proteins, serosal proteins, and cell membrane receptors, generally synthesized by the liver, is found in the blood as inactive precursors. When stimulated by one of several triggers system that can be activated to destroy bacteria. Triggered system can initiate a chain reaction forming antibodies and sensitized lymphocytes to destroy the responsible target. They account for about 5% of the globulin protein fraction of blood serum.

Immunoglobulins (antibodies) are synthesized by B-lymphocytes. These glycoprotein molecules can tag a microorganism, infected cell or chemical-threat agent (antigen) and deactivate them. Antibodies can be non-specific as found in normal blood stream or can be synthesized to destroy one target selectively.

3.3.3 Systemic toxicity and hypersensivity

Of the most important features of biomedical materials non-toxicity, in other terms material must not harm the living organism. Damage of substance to the organism may vary from mild local response to (sometimes) lethal generalized reactions. Acute toxicity can be explained as damaging affects to an organism with a single or short-term exposure. Any natural or artificial substance (animal or plant secretions or extracts, inorganic materials, drugs) may damage the living tissues gradually related with their potency, doses and administration way and metabolism. Even materials like inevitable components of cells, like water, or even oxygen, may be harmful for our organism in higher doses. Pure substances may have toxicity risk solely, correspondingly compound materials, that contain more than one component have a different risk rate for each ingredient. Some components may induce toxicity rate of each other or may act a synergic action to harm organism. The damage to living structures, correspondingly with the amount of substance, may occur in many cascades of internal or external cell functions or structures.

Chronic toxicity is the ability of a substance to cause harmful effects over an extended period (sometimes for the entire life), usually on repeated or continuous exposure. Therefore, corrosion, degradation or decomposition of material placed inside the body is, unless designed to act in this direction, is not desired due to the risk of local or systemic toxicity risk. As the toxicity occurs; depending its damage level, patients' complain may vary from local irritation to life-threatening severe condition clinically until the reason of toxicity is removed or neutralized.

Deactivation process of any molecule sensed as a threat (antigen) by immune system is allergic reaction. Several defence mechanisms are initiated locally and neutralize the antigen. However; sometimes, no matter what the amount of antigen entered to the body it may trigger chain-reaction over the normal limits and carries the reaction via chemical signalling all around whole body and described as hypersensitivity reaction. The patient could have been pre-sensitized by multiple exposures to the antigen until that time. No or mild previous reactions due to the antigen contact can be reported by the patient and sudden onset of hypersensitivity may occur at a severe level. On the other hand; the organism may have a genetic potential to generate this over-reaction, which means hypersensitivity reaction may occur during the first introduction with the antigen material and can be lethal.

3.3.4 Tumorogenesis and biomaterials

Reproduction of healthy cells is division of cells (mitosis). DNA controls the frequency of mitosis, likewise all functions of cell. During the life span of a healthy cell DNA carries a code to end cell life and programmed cell death (apoptosis) occurs. However, some external factors, like chemical agents, radiation, some bacteria or virus infections, may cause a change on the DNA structure (mutation) or pre-determined genetic code is deteriorated during normal life span [75,76]. These mutagenic alterations lead DNA to lose its inhibitory control on mitosis or moreover induce continuous cell division. Additionally or on the other hand; the genetic mutation may clear apoptosis order of the nature and continually dividing cells can survive forever. Both mechanisms cause unstoppable cell number increase inside the body, called as tumourogenesis (tumour formation).Some substances induce cell division via DNA mutation and cause tumour formation, while some, like ethyl alcohol and oestrogen, may promote mitosis and lead cancer. Similarly; continual physical, chemical, electrical or electromagnetic stimuli, behave as chronic irritation and trigger the healing mechanism, which means inducing cell division to repair the injured area. Unless the factor reasoning promoted cell division is relieved cancer out-breaks would be faced. Those are the mainly accepted carcinogenic theories, which give the clues to be considered, while biomaterial production is concerned.

3.3.5 Implant-associated infection

A material or object placed in to an organism to restore a function, mass loss or to measure, diagnose or treat any condition is called as "implant". Implants' success is directly related to their survival. Life expectancy of the implants can be interfered with several endogenous or exogenous factors [77]. Least but not the last issue considering implant failure is the infection. Implant, as a foreign material has a risk to undergo rejection. Therefore; implants are produced of biocompatible materials and sterilized before placement into the body. As antimicrobial body protection is carried out by our skin and mucosal barrier, inner environment of the body is sterile, except some parts like digestive system canal etc. Microorganisms are always in contact with this shield and are not let go inside. However, any kind of injury may break the integrity of the skin-mucosa barrier and can open a gate to the microorganisms to enter body. Following their entry, they replicate themselves and retain at a weak part of organism (colonization). During the proliferation process the also synthesize and release some toxins to damage the host to weaken its defence or to make the environment more proper for their survival. Inoculation of microorganism to the implant material can be via direct contact or blood stream can act like a high-way to transport the microorganisms from a distant entry point to the implant. Either way of exposure can easily initiate colonization of the microorganisms on the implant, in other words infection. Implant, itself, has a risk of rejection and invasion by the microorganisms would lead and hasten withdrawal of the implant. Especially mouth, as a wet and warm atmosphere full of food debris, is an ideal environment for bacteria colonization. That makes it a unique part of the body, with the highest type-multiplicity of microorganism and highest microorganism concentration in one unit of saliva. Those factors mentioned above increase the infection risk of dental implants placed in edentulous jaw bones to restore chewing, phonation functions and aesthetic. Implant surface roughened intentionally to increase bone-implant contact area for more stability and longer survival in the bone cavity. Rough surface provide an ideal environment for bacteria colonization with its retentive topography and can facilitate increase of bacteria count on dental implant. Moreover; incidence of dental implant placement is the higher than all other types of implants. This high frequency of application also increases the infection coincidence of dental implants. Dental implant infections can be as high as 31,2% [78].

Overcoming implant associated infections is another focus of current researches. Previous efforts focused on killing already-colonized microorganisms via several antimicrobial agents like antibiotics [79] or debriding the colonies with several techniques like scratching, chemical cautery or damaging the colonies via photo thermal effect of laser energy. However current concepts have centered on prevention of colonization or may be killing the microorganism as they contact the implant. Those studies showing positive effect of antibiotic use on infection treatment grounded the idea of keeping antibacterial agents around the implant site during healing process. Initial attempts aimed administration of antimicrobials, like tetracycline, ciprofloxacin, vancomycin, rifampin/minocyclin, cefoperazone, penicillin/streptomycin, gentamicin, around the implant to prevent infection. However; those local administration of antibiotics to the implant site topically had limited release time and couldn't prevent development of infection in long term. Idea of gradual release of the antimicrobial substances by the implant has leaded coating titanium surface with antibiotics with restricted success limited time of degradation process of antimicrobial drug coating [80]. Several local drug delivery vehicles like (glycolic, lactic acid, caprolactone, methyl methacrylate polymers, chitosan, agarose/hydro-gels, bioactive ceramics, collagen, nanotubes), have been studied to prolong releasing period of the antibiotic [81-83]. Choice of delivery medium is dependent on:

- Type of tissue to be placed (whether soft tissue or bone)
- Environment of the implant to be protected (mouth, skin, bone, vessels, heart, genitor-urinary system, aero digestive system, cranium or any place inside the body)
- Required releasing time (slow or fast)
- Biodegradation mechanism
- How many drugs will be used (single or multiple) and
- Compatibility with antimicrobial agent to be sustained.

Simultaneously; antimicrobial prevention concepts have evolved innovative surface technologies based on the knowledge that certain metal ions like silver, copper, bismuth and zinc have oligodynamic effect (toxic effect on living organisms) on microorganisms [84,85]. Target of those researches centered on modifying the titanium dental implant surface with these above mentioned metal ions to wall-up a defensive line on the titanium surface and protect themselves against bacterial attacks. Similarly zirconium doped titanium presents antimicrobial activity beside high epithelial cell attraction and mediate healthy cell proliferation to promote formation epithelial cell barrier over the implant surface.

Development of carbon-nanotube systems grown on implant surface can give the capability of controlled release of any drug implanted. It has been shown that they can act as sensing probe for various (electrical, thermal, photo or chemical) stimuli. These triggering factors can be redox reactions of bone-forming cells (osteoblasts) or connective tissue-forming cells (fibroblasts) or any substance specifically released from bacterial wall. As impulses are sensed, such materials can release any given drugs to potentially fight bacterial infection, reduce inflammation, promote bone growth or reduce fibroblast functions [85, 86].

3.3.6 Blood coagulation and blood materials interactions

Any kind of injury may disrupt integrity of blood vessels and cause bleeding. Coagulation (hemostasis) mechanism is to stop blood loss, which can also be named as a protective mechanism. Injury of the vessel initiates physical, cellular and biochemical chain reaction.

Physical change due to trauma (rupture) of vessel walls leads a reflex at surrounding muscles of the vessel and they constrict diameter of vessel, so that blood flow thru narrower lumen decreases gradually and contribute success of co-working hemostatic mechanisms. Besides external trauma, internal biochemical stimulus released by platelets (thromboxane A_2), has an additive effect on vasoconstriction [75,76].

Cellular part of hemostasis mechanism is fulfilled by platelets, formed in the bone marrow and release into blood stream as a member of blood cell population. Platelets, though lack of reproduction capability and nuclei in their cytoplasm, contain thrombosthenin, that contract the platelet cell initially. They can store calcium ions in their cytoplasm and synthesize several molecules act actively in inflammation process (pain mediators).

Main action of platelets in hemostasis mechanism is aggregation via adhesion on another and to injured vessel walls. Glycoprotein layer on cell membrane presents a selective adhesiveness to the platelet. Under normal conditions this layer rebuff adhesion to healthy vessel wall tissue (endothelial cells), while in presence of any injury glycoprotein promotes adhesion to the damaged endothelium, collagen bundles supporting vascular wall and one another. This increased adherence occurs simultaneously with changing their form as swelling and presenting multiple protrusions on their cell membrane and degradation of several substances of some attract more platelets and lead increase of those cells in number. Increased irregularity of their form and sticky nature of the membrane ease their adhesion capability to accumulate at the damaged vessel window. These aggregation and adhesion behavior of platelets yield initial clot formation to stop bleeding.

The third method playing a role in hemostasis mechanism simultaneously with vascular response and platelet plug formation is biochemical coagulation cascade. Basically; over 50 substances found in blood or tissue take place in blood coagulation (procoagulants) or anticoagulation (anticoagulants) equilibrium. Procoagulant factors of those substances (Table 17) are present in inactive (precursor) form and any stimulus originated traumatized vascular endothelium or platelet derivates can activate them to start the coagulation mechanism. These extrinsic and intrinsic pathways activate those factors and each activated substance either activates another or catalyse their activation process. Final steps of the cascade are change of fibrinogen to fibrin via thrombin and formation of fibrin network as an end product.

Factor I	fibrinogen
Factor II	prothrombin
Factor III	tissue factor or tissue thromboplastin
Factor IV	calcium ion
Factor V	proaccelerin, labile factor or Ac-globulin
Factor VII	serum prothrombin conversion accelerator, proconvertin or stable factor
Factor VIII	antihemophilic factor, antihemophilic globulin, antihemophilic factor A
Factor IX	plasma thromboplastin component, Christmas factor or antihemophilic factor B
Factor X	Stuart factor or Stuart-Prower factor
Factor XI	plasma thromboplastin antecedent or antihemophilic factor C
Factor XII	Hageman factor
Factor XIII	fibrin-stabilizing factor
Fletcher factor	Prekallikrein
Fitzgerald factor	High-molecular-weight kininogen
Platelets	thrombocytes

Table 17. Factors in blood coagulation mechanism

Blood clot (thrombus) piece can be detached from a main clot mass or sometimes may occur spontaneously without any injury effect. High blood viscosity, high procoagulant serum levels or vascular malformations etc. increase intravascular thrombus formation risk. Any free thrombus piece can easily be transferred via circulatory system until get stacked in a narrower vessel lumen and obliterate an artery (of heart, brain, kidney eye etc.), which may cause a severe damage due to occluded blood circulation of the related the tissue. Especially, thrombus formation can easily occur on irregular vascular surfaces covered with cholesterol plaques or around any vascular implants or grafts like stents used in cardiovascular interventions. For a healthy person; anticoagulation mechanism is a protective equilibrium that impedes coagulation of whole blood in the vascular system spontaneously, which may mean end of life. In other case, as the clotting cascade begins due to any reason, anticoagulation mechanism obviates dissemination of the coagulation chain thru whole circulation system. Anticoagulant substances found in our body like vascular surface cover, glycocalyx, helping platelet or clotting factor repelling, or thrombin binding protein thrombomodulin. Fibrin fibers formed during coagulation consumes thrombin and eliminates it from the environment. Additionally alpha globulin protein (antithrombin-heparin cofactor or antithrombin III) also removes thrombin and hinder or delay coagulation. In our daily life; anticoagulation is an equilibrium is maintained via vascular surface integrity, presence of those anticoagulant substances in the blood or with some drugs.

On the other hand, in case of lack of any of those clotting factors or a deficit in platelet structure, function or count lead disturbed coagulation function, which means unstoppable bleeding reasoning death. Anticoagulation kept in normal limits is desired to avoid intravascular coagulation especially in patients with vascular malformations, vascular grafts or stents, atherosclerosis etc. Therefore, those patients are prescribed anticoagulant, like heparin, coumarin or 1,3-indandionederivates, to increase coagulation threshold within safety limits.

3.4 Evaluation of materials behavior

Prior to clinical use of any substance (biomaterial or drug), which is going to be in contact with our organism, it should be strictly tested and proven to be non-hazardous for us. Those processes to examine this biomaterial can be classified in 3 parts. First requirement is the matching physical and chemical properties of the substance, desired to be the same or at least compatible with the organism. These materials, therefore, are produced according the tissue features, where they are going to be used. According to tissue type; physical strengths (tensile, compressive, shear strength, elasticity modulus), thermal properties, photoreactivity or translucency, colour, calcification potency, surface structure, chemical features, or degradation resistance are modified for ideal adaptation to the environment. Those features are examined under laboratory conditions before biologic behavioral tests [87].

3.4.1 In vitro assessment of tissue compatibility

Proofed materials in laboratory examinations are ready to be evaluated for their biologic performance. According to the tissue(s) which is(are) going to be the environment of the material, related tissue cells are selected to evaluate if the material has a potential to damage them, or not. Even materials previously proven to be safe to be use inside the body, should be tested if any chemical or major physical modification has been carried out. With this aim cell culture systems are the first step evaluation process to assess toxicity of the material for related cell types. In vitro term is used to define a test setup that produces cells extracted from a living organism outside the body in controlled laboratory conditions. This method is helpful to assess the biologic behavior of a material without killing plenty of animals used in experiments. Additionally, these in vitro study settings result rapidly, can be performed in many different models and can be repeated in a short time. Initially cells are harvested from a living organism (animal, human or even plants), kept in proper environmental conditions containing all necessary organic and inorganic substances, water, temperature etc., to maintain their survival. They should be isolated from other cell groups or microorganisms to gain a pure cell population with equivalent features, so that cellular response could be same in this population. Genetic structure can be modified to give them immortality to keep their production continuously. Gene modification can be used to transform those cells to cancer cells to work on such illnesses. Then cultured cells are passaged and exposed to test material or can be stored at -196°C for further studies. Time depended toxicity rate can be measured as the count of surviving cells, their physical appearance, functions like replication rate or energy production ability (mitochondrial activity) [88].

Additionally, biologic behavior of a material is not only toxicity but also can be an altering effect on cell structure. Especially mutation effect on cell DNA due to test material exposure may also change cellular behavior or character and cause cancer. Therefore, such mutagenity evaluations can also be made for further investigation. With this purpose morphological change can be scanned with scanning electron microscopy or ultrastructural transformations can be visualized with transmission electron microscopy. Cellular affinity to the substance is eventually another point of interest to measure. Cellular extracts are lined on the prepared surface and incubated for a certain period to wait for cell adhesion. Then cultured specimens are washed out with water and examined to count attached cell numbers.

Some behavioral and structural alterations can be monitored with various biochemicals, spectrophotometric, immunoassay techniques (ELISA-Enzyme-Linked Immunosorbent Assay) or dynamic imaging methods like confocal microscopy. These techniques can detect presence of a defined substance with various methodology in a sampled moment or in real-time.

On the other hand, some materials like bone grafts are used as a scaffold for tissue engineering procedures that mainly act as a carrier of implanted cells and biomolecules. In advanced applications; differentiation of root (stem) cells planted in/on those graft materials to bone cells is another desired feature expected from a bone graft material. Hence, differentiation effect of test material can also be examined with undifferentiated stem cells.

Besides cell cultures; bacterial or fungi cultures can also be used to evaluate antimicrobial effect of the test material. On microorganisms cultured at a medium (on a surface or in a solution) are exposed to the test material and surviving microorganisms are counted to evaluate their lethal efficacy [89,90].

3.4.2 In vivo assessment of tissue compatibility and animal models

Third step of biologic behavioral tests are in vivo (animal) experiments. Only approved substances with in vitro examinations are subjected to in vivo experiments, so that excessive scarification of experiment subjects is minimized. Depending on the purpose of biomaterial; the substance is tested whether it cause any allergic reaction (sensitization), tissue inflammation and rejection reaction (irritation). It is also implanted under the skin (intracutaneously) or in a prepared bone cavity (intrabony) to evaluate local tissue response. Additionally following administration in to the body (no matter what way of administration was chosen) acute or subacute damages to all organ systems (systemic toxicity) are evaluated after implantation via examining each organ with microscopic imaging or biochemical analyses. Genotoxicity potential of the studied material is investigated with changes of genetic structure of the cells exposed to the material. Changes of the healthy tissue can be assessed with comparison of those tissue samples exposed to the substance with healthy tissue structures of the same spice with different methodology. Tissue morphology can be evaluated with microscopic techniques like light microscopy, scanning electron microscopy, transmission electron microscopy, confocal microscopy etc. or some physical tests (tensile strength, shear strength or compression tests) can be used to evaluate physical strength of hard tissues like bone and tooth, or even (sometimes) soft tissues. Biochemical alterations, before and after application of the researched material, can also be compared to evaluate tissue response or its reaction path to the material. Additionally calcified tissues like bone or tooth can be followed up via X-ray images (radiographically). Study designs can be based on time dependent comparative research methodology and can use assessment of half-life of radioactively singed atoms implanted to the study substance to be placed in to the body [91].

Samples to be examined should be free of any contamination (pure), packed properly and sterilized according their structure (with gamma-radiation, steam+heat, ethylene oxide or plasma). Local reaction to the material can be measured with comparison of healthy tissue, physiologic tissue healing or healing process of a standard material well-known and documented previously. In such methodology, unless systemic efficacy is not in consideration, more than one substance can be tested in one organism to make a healthy comparison. However, at least 6 subjects should be studied for each group for statistical analysis. If time depended results are required one group of animals should be prepared for each time interval. Examinations regarding systemic effect of the substance necessitate use of one material on one animal each time, which eventually increase number of animals. Regarding systemic toxicity, irritation or hypersensitivity repetitive dose applications can be used, which also means raised animal number. Following application of test material, the subjects are kept in a convenient habitation conditions. Responses are monitored after scarification or real-time with suitable probing in tissues and body fluids histologically, radiologically, chemically, thermally or immunohistochemically etc. For evaluation of results; obtained raw data is analysed statistically with parametric and nonparametric tests.

Physical or chemical tests can be performed to evaluate structural, chemical or physical strength changes of the material itself after using in the organism. Especially force bearing materials like bone plates or metallic prosthesis or dental implants can be examined regarding corrosion potential, micro/macro cracks or plastic deformation, so that survival of material can be estimated under compressive, tensile and shear strengths besides corrosive effect of body fluids [92].

3.5 Dental implants

Other than restorative dentistry the rehabilitation of teeth losses is a major concern for clinicians. Replacing the teeth with human-made one has been one of the greatest challenges in field of dental medicine. Mechanical features like macro anatomy, micro surface topography of dental implants necessitate a special concern to give a life-long service in an environment where repeating bite or chewing forces are loaded from different directions. These physical design properties are expected to neutralize those loaded chewing forces or at least should minimize compressive stresses on the bony bed where they were implanted. Moreover; regarding resistance against metal fatigue choice of material is highly important in case of cycling compressive, shear and tensile loads. Dental implants are partly placed into the jaw bones where in contact with blood and also have a part left in the oral cavity where exposed to saliva. Those body fluids that have lipophilic and hydrophilic affinity and enzymatic activity potentially dissolve and corrode materials. Hence corrosion resistance of the dental implant material should be high enough both to sustain their physical strength in acceptable limits and also to minimize local or systemic toxicity risk due to corrosion particles. However; first place for significance ranking of dental implant material, likewise in all biomaterials, is its compatibility with the surrounding tissues (biocompatibility). Therefore; several materials and alloys have been tried for decades to find out the ideal biocompatibility with excellent physical features. Titanium and zirconium elements have been well documented and presented to have better tissue biocompatibilities and acceptable physical properties, when compared to others [93,94]. It is been shown that alloys of these materials can present augmented biologic and physical features while corrosion aspect of those alloys is still questionable. With this regard Ti-6Al-4V, Ti-6Al-7Nb, Ti-5Al-2Nb-1Ta, Ti-30Ta and Ti-Zr alloys have been studied in several studies to enhance mechanical and biologic properties of titanium or zirconium [95-99]. Those alloys have been used as base material or as coating for augmented features. More recent studies focusing on Ti-Zr alloy have promising results which show evidences of elimination of disadvantages of both substances while improving biocompatibility.

Except using alloys to improve biocompatibility several techniques have been used to modify implant surface that is in contact with bone. Changing micro-structural, chemical or ionic structure of dental implant material to make it more attractive for surrounding bone cells (osteoblast) and allow it to bind calcium and phosphate ions of neighboring bony bed [100].

Following preparation of a place in the jaw bone dental implant is placed into the cavity and left for healing. During the healing process blood clot cover implant surface which is not in close contact with bone. Those two different neighboring structures (bone and blood clot) initiate healing process which means integration of implant to surrounding bone tightly. Parts of implant surface at direct bony contact may develop chemical connection with calcium and phosphate of bony hydroxyapatite crystal. Developing a bone contact on the implant surface exposed to blood clot needs more time to have bony connection. Overlaying clot undergoes transformation turn into bone tissue, so that implant connects to surrounding bone (osteointegration). Both titanium and zirconium materials and their alloys are capable to show osteointegration. Considering this process, implant surface must be available for osteoblast habitation, in other terms implant surface should have an optimum surface roughness values to let the bone cells attach on tightly. Surface treatment also helps to increase surface magnitude with roughening, which also contribute extension of bone-to-implant contact area. Following adhesion; cells synthesize osteoid bone matrix (immature bone) that acts like a template leading calcium hydroxyl apatite crystal precipitation in (calcification). Therefore, osteoblastic attachment is inevitable for osteointegration, which also means "success" [101]. Therefore; increasing surface roughness has been worked out with different techniques, like sand blasting, acid etching, abrading with SiC paper, microarc oxidation, spark erosion, lasering or their combinations [102,103]. Sand blasting is a well-documented method to improve surface porosity, as well as altering several physical properties of the metal (fracture, fatigue, tensile strength etc.) [104].

Various parameters have been modified for optimization both for roughness and bone cell (osteoblast) adhesion to the surface. Different materials like aluminium oxide, TiO_2, hydroxyapatite, calcium phosphate and zirconium oxide have been studied in different particle sizes in combination or separately [105]. Parameters like blasting speed, particle diameter, and particle ratio in the blowing air, blasting atmosphere and temperature can affect surface topography. Type of blasting material is also important with the respect that during the blasting process some particles can be stacked into the titanium surface due to high velocity and may remain on even after cleaning processes following blasting. It should be considered that those blasting particles remaining on the implant surface can alter tissue reaction to the implant material in positive or negative direction. With this regard, choice of blasting materials among biocompatible ones would accelerate osteointegration process, while substances with low or no biocompatibility can be expected to influence this process negatively [106]. It should also be considered that these blasting materials remaining on the implant surface may undergo corrosion and may result rejection of the implant or toxicity by the time as well [107]. Therefore, analysis of the titanium surface following blasting can be advised to evaluate any remnants of blasting materials if such susceptible material had been used for blasting. Owing that; several chemicals or their combinations have been tried out and some are currently used to modify implant surface to enhance roughness. Phosphoric acid and its derivates, HF, HNO_3, HCl and H_2SO_4 the most frequently used chemical etching agents to modify titanium surface solely or in combination. It is also noted that temperature is an important factor that increase the corrosive effect of chemicals [108]. Even though several research groups showed that acid treatment is more useful then alkali solutions to improve surface roughness of titanium [109], there are studies showing that alkali solutions are efficient for surface treatment of titanium implants in nanoscale and moreover can be advantageous to give hydrophilic structure on the implant surface, which provide augmented tissue affinity to the titanium surface [110]. It is also been stated that alkali- and heat-treated titanium surface can accelerate and improve bone-implant contact area due to increased surface roughness [111]. As understood from those results; alkali treatment can be attributed as less aggressive than acid etching method, but can be used following acid treatment and forms nanoroughness on the titanium surface that can attract bone cells to attach on. Hydrogen peroxide, likewise alkali treatment, and heat treatment can improve cell adhesion and create hydrophilic nature for titanium [112]. Moreover; alkali (NaOH) treated titanium surfaces can induce biomimetic apatite debris formation on the implant surface, which can be attributed as chemical binding of titanium and bony apatite. This bone-like apatite focuses act as bone calcification nuclei and initiate bone formation.

Surface treatment with previously mentioned methods, sand blasting and acid or alkali etching, do contaminate the titanium surface, of which remnants cannot be avoided, eradicated or neutralized totally all the time. However, laser energy can modify the titanium or titanium alloy surfaces without contamination [113]. Laser energy can change the titanium or its alloys with heat (photothermal) effect or intense pulsed wave form of laser energy (photomechanic) effect like embossing. For this purpose, certain wavelengths, like neodymium: yttrium aluminium garnet (Nd-YAG) laser-355 nm and carbon dioxide laser-10600 nm, have been well documented [114] and taken an inevitable place in industrial surface treatment industry. Pulsed Nd-YAG laser with 10 Hz repetition rate has been demonstrated to give the opportunity to control micro-topography as desired [115]. However, those wavelengths work with photothermal effect and modify the titanium surface via melting. Even laser surface treatment is defined to give the control for desired surface topography melting method does not create a perfect surface structure. Therefore, shorter (femto or pico second) pulses can generate more controlled surface shape due to lower heat formation [116]. Femtosecond laser pulses can engrave surface topography at micro or nano scales, which give the 100% control facility of surface texturing. Control of roughness depth and geometry on the implant surface also gives the possibility for selective cell attraction to the implant surface. Such laser-treated surfaces can distract inflammatory cells, responsible from tissue response to implant surface, and contribute to suppression of early inflammatory events, which may lead rejection [117]. Similarly, laser-texturing can make it possible to attract certain cells of which long-term function is a prerequisite for their adhesion, like epithelial cells and osteoblasts. Those cells attach directly to the implant surface, if convenient roughness values can be yielded with treatment. Connective tissue cells (fibroblast) tend to attach less roughened surfaces, when compared with bone cells (osteoblasts) [118]. Wavelengths proven to be innocuous to the titanium surface like erbium: yttrium aluminium garnet (Er:YAG) laser-2940 nm, recently, have been documented as a laser type that may modify the titanium surface at certain power settings as low as 200 mJ/10Hz [119]. On the other hand, degree of surface roughness can alter surface adhesion potential of several microorganisms as well as body cells [120]. Owing that affinity potential, implant surfaces can be modified gradually according to the location where cell adhesion is targeted, while microbial attacks are repulsed.

Titanium surfaces treated with micro-arc (plasma electrolytic) oxidation (MAO) can also improve surface porosity and its alloys, contributing cellular activity for osteointegration [121]. Correspondingly; MAO procedure facilitates cell adhesion capability of titanium surface with enhanced hydrophilicity [122].

4. Summary

In conclusion, physical, chemical and biologic features of biomaterials are chosen and determined considering their function and required durability in situ besides biocompatibility. Accelerative development in material science, especially in nano and optoelectric sciences makes exploration of new materials or enhancement of conservative biomaterials.

5. Acknowledgements

The authors would like to thank Dr. Kadriye Atıcı Kızılbey and Chem. Çağdaş Büyükpınar from Yıldız Technical University, Science and Technology Application and Research Center (Turkey) for their kindly assistance in arrangements of the present chapter.

6. References

[1] Boretos JW, Eden M (1984) Contemporary Biomaterials, Material and Host Response, Clinical Applications, New Technology and Legal Aspects. Noyes Publications, Park Ridge, NJ, pp. 232–233.

[2] Williams DF (1987) Review: Tissue-biomaterial interactions. J. Mat. Sci. 22 (10): 3421-3445.

[3] http://users.ox.ac.uk/~exet0249/biomaterials.html#biomat

[4] Niinomi M (2002) Recent Metallic Materials for Biomedical Applications. Metal. Mater. Transac. A. 33 A: 477-486.

[5] http://www.biomedicalalloys.com/home.html

[6] http://media.wiley.com/product_data/excerpt/44/04712539/0471253944.pdf

[7] Hermawan H, Ramdan D, Djuansjah J R P (2011) Biomedical Engineering – From Theory to Applications. In: Reza Fazel-Rezai, editor. Metals for Biomedical Applications. Rijeka: InTech. pp. 411-430.

[8] Manivasagam G, Dhinasekaran D, Rajamanickam A (2010) Biomedical Implants: Corrosion and its Prevention - A Review. Recent Patents on Corrosion Science. Vol. 2. pp. 40-54.

[9] Park J P, Bronzino JD (2003) Biomaterials: Principles and Applications. In: Kon Kim Y, Park JB, editors. Metallic Biomaterials. USA: CRC Press LLC. pp. 1-20.

[10] Chandra R, Rustgi R (1998) Biodegradable Polymers. Progress in Polymer Science, 23: 1273.

[11] http://upload.wikimedia.org/wikipedia/commons/6/64/Protein_TF_PDB_1a8e.png

[12] https://chempolymerproject.wikispaces.com/file/view/DNA.gif/34197899/374x345/DNA.gif

[13] http://www.wikidoc.org/index.php/File:Protein-primary-structure.png

[14] Van der Rest M (1991) Collagen Family of Proteins. The FASEB Journal. 5: 2814-2823.

[15] http://thegist.dermagist.com/wp-content/uploads/2011/01/collagen.jpg

[16] Lawton J.W (2001) Zein: A History of Processing and Use. Cereal Chem. 79(1):1–18.

[17] Varki A, Cummings R, Esko J, Freeze H, Stanley P, Bertozzi C, Hart G, Etzler M (2008) Essentials of glycobiology. Cold Spring Harbor Laboratory Press. 2nd edition.

[18] http://independent.academia.edu/PaulMuljadi/Teaching/30666/Cellulose

[19] Ravi Kumar MNV (2000) A review of Chitin and Chitosan Applications. Reactive and Functional Polymers. 46(1): 1-27.

[20] Kittel C (1996) Chapter 1: Introduction to Solid State Physics (Seventh ed.). New York: John Wiley & Sons. pp. 10.

[21] http://commons.wikimedia.org/wiki/File:Point_defects_in_crystal_structures.svg

[22] http://www.substech.com/dokuwiki/doku.php?id=imperfections_of_crystal_structure

[23] http://moisespinedacaf.blogspot.com/2010/06/planar-defects-and-boundaries.html

[24] Pilliar RM (2009) Metallic Biomaterials. In: R. Narayan, Editor. Biomedical Materials. Springer Science+Business Media LLC. Chapter 2. pp. 1-42.

[25] http://www.uweb.engr.washington.edu/education/Bioe599/Hoffman2.pdf

[26] http://medgadget.com/2012/02/nanodiamond-toughened-orthopedic-implants-show-promise-in-study.html

[27] http://www.brmb.co.uk/news/headlines/fears-over-hip-replacement-poisoning/

[28] http://www.worldofstock.com/stock-photos/xray-after-osteosynthesis-metal-implants-to-adjust/PHE3885

[29] http://www.lifescript.com/health/a-z/treatments_a-z/procedures/c/coronary_stenting.aspx

[30] Silver F H, Christiansen D L (1999) Biomaterials Science and Biocompatibility, Springer-Verlag 87-120.

[31] Schieker M, Seitz H, Drosse I, Seitz S, Mutschler W (2006) Biomaterials as scaffold for bone tissue engineering. European Journal of Trauma, 32(2): p. 114-124.

[32] Hutmacher DW, Sittinger M, Risbud MV (2004) Scaffold-based tissue engineering: rationale for computer-aided design and solid free-form fabrication systems. Trends in Biotechnology, Vol. 22(No.7): p. 354-362.

[33] Hutmacher DW (2000) Scaffolds in tissue engineering bone and cartilage. Biomaterials, 21: p. 2529-2543.

[34] Gomes ME, Godinho JS, Tchalamov D, Cunha AM, Reis RL (2002) Alternative tissue engineering scaffolds based on starch: processing methodologies, morphology, degradation and mechanical properties. Materials Science and Engineering C, 20: p. 19-26.

[35] Anselme K (2000) Osteoblast adhesion on biomaterials. Review. Biomaterials, 21: p. 667-681.

[36] William J O'Brien, PhD, FADM (Ed.) Dental Materials and Their Selection Fourth Edition Quintessence Publishing Co, Inc.

[37] R Williams (Ed.) (2010) Surface modification of biomaterials: Methods analysis and applications, Woodhead publication Limited Cambridge.

[38] Vo-Dinh T, Cullum B (2000) Biosensors and biochips: advances in biological and medical diagnostics Fresenius J Anal. Chem. 366:540–551.

[39] Rossen L, Nørskov P, Holmstrøm K, Rasmussen OF (1992) Inhibition of PCR by components of food samples, microbial diagnostic assays and DNA-extraction solutions International Journal of Food Microbiology   Volume 17, Issue 1, Pages 37-45.

[40] Marvin Ryou MD, Christopher C, Thompson MD (2006) Tissue Adhesives: A Review Techniques in Gastrointestinal Endoscopy Volume 8, Issue 1, pp. 33-37.

[41] Mikos AG, Bao Y, Cima LG, Ingber DE, Vacanti JP, Langer R (1993) Preparation of Poly(glycolic acid) bonded fiber structures for cell attachment and transplantation. Journal of Biomedical Materials Research, 27: p. 183-189.

[42] http://repositorium.sdum.uminho.pt/bitstream/1822/7617/1/MasterThesis_Novelpolymericsystems_SLuna.pdf

[43] Nair LS, Laurencin CT (2006) Polymers as Biomaterials for Tissue Engineering and Controlled Drug Delivery. Adv Biochem Engin/Biotechnol 102: 47–90.

[44] Strobl G (2007) The Physics of Polymers: Concepts for Understanding Their Structures and Behavior. Third Edition. Springer-Verlag: Berlin Heidelberg.

[45] http://www.google.com.tr/imgres?q=styrene+monomer&um=1&hl=tr&lr=&biw=1280&bih=904&tbm=isch&tbnid=3lG4on1irwW1lM:&imgrefurl=http://pslc.ws/macrog/kidsmac/polysty.htm&docid=WucdRJBNPTnr1M&imgurl=http://pslc.ws/macrog/kidsmac/images/styrene.gif&w=118&h=143&ei=0GJXT4KAJOmh4gTZrvCbDw&zoom=1

[46] Ebewele RO (2000) Polymer Science and Technology, CRC Press LLC.

[47] http://www.dowcorning.com/content/publishedlit/01-1112-01.pdf

[48] Kamal MR, Huang B (1992) Natural and artificial weathering of polymers. In Hamid, S.H., M. B. Ami, and A. G. Maadhan. Eds., Handbook of Polymer Degradation. Marcel Dekker, New York, NY pp. 127-168.

[49] Tian H, Tang Z, Zhuang X, Chen X, Jing X (2012) Biodegradable synthetic polymers: Preparation, functionalization and biomedical application Progress in Polymer Science 37 pp. 237– 280.

[50] Messer RL, Lockwood PE, Wataha JC, Lewis JB, Norris S, Bouillaguet S (2003) In vitro cytotoxicity of traditional versus contemporary dental ceramics. Jounal of Prosthetic Dentistry 90: 452-458.

[51] Yamamoto A, Honma R, Sumita M, Hanawa T (2004) Cytotoxicity evaluation of ceramic particles of different sizes and shapes. Journal of Biomedical Materials Research Part A Volume: 68, Issue: 2, Pages: 244-256.

[52] Thamaraiselvi TV, Rajeswari S (2004) Biological Evaluation of Bioceramic Materials - A Review. Trends Biomater. Artif. Organs, Vol 18 (1), pp 9-17.

[53] Billotte, W G (2000) The Biomedical Engineering Handbook: Second Edition. In: Joseph D. Ed. Chapter 38: Ceramic Biomaterials. Bronzino Boca Raton: CRC Press LLC.

[54] Vallet-Regí M (2001) Ceramics for medical applications. J. Chem. Soc., Dalton Trans., 97–108.

[55] http://earthsci.org/mineral/rockmin/chart/nacl.gif

[56] http://www.metafysica.nl/turing/preparation_3dim_3.html

[57] http://www.csa.com/discoveryguides/archives/bceramics.php

[58] Dorozhkin SV (2010) Bioceramics of calcium orthophosphates Biomaterials 31: 1465– 1485.

[59] http://www.oxforddentalcentre.co.uk/Implants.html

[60] http://www.aap.de/en/Produkte/Orthobiologie/Knochenersatz/Cerabone/index_html

[61] http://www.phoeniximplantdentist.com/zirconia.html

[62] http://www.ioi.com/

[63] Dorozhkin SV (2009) Calcium Orthophosphates in Nature Review Biology and Medicine Materials. 2: 399-498.

[64] http://jba.sagepub.com/content/23/3/197.full.pdf+html

[65] Iftekhar A (2004) Standard Handbook of Biomedical Engineering and Design. Chapter 12: Biomedical Composites McGraw-Hill Companies.

[66] http://imeulia.blogspot.com/2011/08/classes-and-characteristics-of.html

[67] Bronzino JD (Ed.) (2000) The Biomedical Engineering Handbook: Second Edition Lakes, R. Composite Biomaterials. Boca Raton: CRC Press LLC.

[68] Ramakrishnaa S, Mayer J, Wintermantel E, Leong KW (2001) Biomedical applications of polymer-composite materials: a review Composites Science and Technology 61: 1189-1224.

[69] Wang M (2003) Developing bioactive composite materials for tissue replacement Biomaterials 24: 2133–2151.

[70] http://massasoit-bio.net/courses/201/201_content/topicdir/skeletal/skeletal_RG/ skeletal_RG4/skeletal_RG4.html

[71] http://www.eolss.net/Sample-Chapters/C05/E6-171-07-00.pdf

[72] http://ecourses.vtu.ac.in/nptel/courses/Webcourse-contents/IISc-BANG/Composite %20Materials/Learning%20material%20-%20composite%20material.pdf

[73] Ashammakhi N, Reis R, Chiellini F (Eds.) (2008) Topics in Tissue Engineering, Chen Q, Roether JA and Boccaccini AR Chapter 6: Tissue Engineering Scaffolds from Bioactive Glass and Composite Materials.

[74] Scholz M-S, Blanchfield JP, Bloom LD, Coburn BH, Elkington M, Fuller JD, Gilbert ME, Muflahi SA, Pernice MF, Rae SI, Trevarthen JA, White SC, Weaver PM, Bond IP

(2011) The use of composite materials in modern orthopaedic medicine and prosthetic devices: A review Composites Science and Technology 71: 1791–1803.

[75] Berne RM, Levy MN, Koeppen BM, Stanton BA (2009) Berne and Levy Physiology. 6th Ed. St. Louis: Mosby.

[76] Guyton AC, Hall JE (2010) Textbook of medical physiology. 12th Ed. Philadelphia: Elsevier Saunders.

[77] Costa FO, Takenaka-Martinez S, Cota LO, Ferreira SD, Silva GL, Costa JE (2012) Peri-implant disease in subjects with and without preventive maintenance: a 5-year follow-up. J Clin Periodontol. 39(2):173-181.

[78] Esposito M, Grusovin MG, Loli V, Coulthard P, Worthington HV (2010) Does antibiotic prophylaxis at implant placement decrease early implant failures? A Cochrane systematic review. Eur J Oral Implantol. (2):101-110.

[79] Karaky AE, Sawair FA, Al-Karadsheh OA, Eimar HA, Algarugly SA, Baqain ZH (2011) Antibiotic prophylaxis and early dental implant failure: a quasi-randomcontrolled clinical trial. Eur J Oral Implantol. 4(1):31-38.

[80] Sirivisoot S, Pareta R, Webster TJ (2011) Electrically controlled drug release from nanostructured polypyrrole coated on titanium. Nanotechnology. 25;22(8):085101.

[81] Gulati K, Ramakrishnan S, Aw MS, Atkins GJ, Findlay DM, Losic D (2012) Biocompatible polymer coating of titania nanotube arrays for improved drug elution and osteoblast adhesion. Acta Biomater. 8(1):449-456.

[82] Silvestri L, van Saene HK, Parodi PC (2011) Decolonization strategies to control Staphylococcus aureus infections in breast implant surgery. Plast Reconstr Surg. 128(1):328-329.

[83] Bumgardner JD, Adatrow P, Haggard WO, Norowski PA (2011) Emerging antibacterial biomaterial strategies for the prevention of peri-implant inflammatory diseases. Int J Oral Maxillofac Implants. 26(3):553-560.

[84] Zhao L, Wang H, Huo K, Cui L, Zhang W, Ni H, Zhang Y, Wu Z, Chu PK (2011) Antibacterial nano-structured titania coating incorporated with silver nanoparticles. Biomaterials. 32(24):5706-5716.

[85] Chang YY, Lai CH, Hsu JT, Tang CH, Liao WC, Huang HL (2012) Antibacterial properties and human gingival fibroblast cell compatibility of TiO_2/Ag compound coatings and ZnO films on titanium-based material. Clin Oral Investig. 16(1):95-100.

[86] Wolf J, Sternberg K, Behrend D, Schmitz KP, von Schwanewede H (2009) Drug release of coated dental implant neck region to improve tissue integration. Biomed Tech (Berl). 54(4):219-227.

[87] Davis JR (2003) Handbook of Materials for Medical Devices. USA. ASM International.

[88] MacGregora JT, Collinsa JM, Sugiyamab Y, et al. (2001) In Vitro Human Tissue Models in Risk Assessment: Report of a Consensus-Building Workshop. Toxicol. Sci. 59 (1): 17-36.

[89] Lönnroth EC, Dahl JE (2001) Cytotoxicity of dental glass ionomers evaluated using dimethyl thiazoldiphenyltetrazolium and neutral red tests. ActaOdontol Scand 59(1):34-39.

[90] Cory AH, Owen TC, Barltrop JA, Cory JG (1991) Use of an aqueous soluble tetrazolium/formazan assay for cell growth assays in culture. Cancer Commun. 3(7):207-212.

[91] Gartner LP, Hiatt JL, Strum JM (2010) Cell Biology and Histology. Baltimore, Lippincott Williams & Wilkins.

[92] Yoruc ABH, Gulay O, Sener BC (2007) Examination of the properties of Ti-Al-4V based plates after oral and maxillofacial application. J OptoelecAdv Mat. 8(9):2627-2633.

[93] Gottlow J, Dard M, Kjellson F, Obrecht M, Sennerby L. (2010) Evaluation of a New Titanium-Zirconium Dental Implant: A Biomechanical and Histological Comparative Study in the Mini Pig. Clin Implant Dent Relat Res. [Epub ahead of print].

[94] Barter S, Stone P, Brägger U (2011) A pilot study to evaluate the success and survival rate of titanium-zirconium implants in partially edentulous patients: results after 24 months of follow-up. Clin Oral Implants Res. [Epub ahead of print].

[95] Gill P, Munroe N, Pulletikurthi C, Pandya S, Haider W (2011) Effect of Manufacturing Process on the Biocompatibility and Mechanical Properties of Ti-30Ta Alloy. J Mater Eng Perform. 20(4):819-823.

[96] Spriano S, Bronzoni M, Vernè E, Maina G, Bergo V, Windler M (2005) Characterization of surface modified Ti-6Al-7Nb alloy. J Mater Sci Mater Med.16(4):301-312.

[97] Tamilselvi S, Raghavendran HB, Srinivasan P, Rajendran N (2009) In vitro and in vivo studies of alkali- and heat-treated Ti-6Al-7Nb and Ti-5Al-2Nb-1Ta alloys for orthopedic implants. J Biomed Mater Res A. 90(2):380-386.

[98] Shapira L, Klinger A, Tadir A, Wilensky A, Halabi A (2009) Effect of a niobium-containing titanium alloy on osteoblast behavior in culture. Clin Oral Implants Res. 20(6):578-582.

[99] Ferraris S, Spriano S, Bianchi CL, Cassinelli C, Vernè E (2011) Surface modification of Ti-6Al-4 V alloy for biomineralization and specific biological response: part II, alkaline phosphatase grafting. J Mater Sci Mater Med. 22(8):1835-1842.

[100] Escada AL, Machado JP, Schneider SG, Rezende MC, Claro AP (2011) Biomimetic calcium phosphate coating on Ti-7.5Mo alloy for dental application. J Mater Sci Mater Med. 22(11):2457-2465.

[101] Anselme K, Linez P, Bigerelle M, Le Maguer D, Le Maguer A, Hardouin P, Hildebrand HF, Iost A, Leroy JM (2000) The relative influence of the topography and chemistry of TiAl6V4 surfaces on osteoblastic cell behaviour. Biomaterials. 21(15):1567-1577.

[102] Wennerberg A, Hallgren C, Johansson C, Sawase T, Lausmaa J (1997) Surface characterization and biological evaluation of spark-eroded surfaces. J Mater Sci Mater Med. 8(12):757-763.

[103] Wieland M, Textor M, Chehroudi B, Brunette DM (2005) Synergistic interaction of topographic features in the production of bone-like nodules on Ti surfaces by rat osteoblasts. Biomaterials. 26(10):1119-1130.

[104] Gil FJ, Planell JA, Padrós A (2002) Fracture and fatigue behavior of shot-blasted titanium dental implants. Implant Dent. 11(1):28-32.

[105] Rønold HJ, Lyngstadaas SP, Ellingsen JE. (2003) A study on the effect of dual blasting with TiO2 on titanium implant surfaces on functional attachment in bone. J Biomed Mater Res A. 67(2):524-530.

[106] Aparicio C, Manero JM, Conde F, Pegueroles M, Planell JA, Vallet-Regí M, Gil FJ (2007) Acceleration of apatite nucleation on microrough bioactive titanium for bone-replacing implants. J Biomed Mater Res A 82(3):521-529.

[107] Aparicio C, Gil FJ, Fonseca C, Barbosa M, Planell JA (2003) Corrosion behaviour of commercially pure titanium shot blasted with different materials and sizes of shot particles for dental implant applications. Biomaterials. 24(2):263-273.

[108] Mizoguchi T, Ishii H (1979) Analytical applications of condensed phosphoric acid-II: determination of aluminium, iron and titanium in bauxites after decomposition with condensed phosphoric acid. Talanta. 26(1):33-39.

[109] Yamaguchi S, Takadama H, Matsushita T, Nakamura T, Kokubo T (2011) Preparation of bioactive Ti-15Zr-4Nb-4Ta alloy from HCl and heat treatments after an NaOH treatment. J Biomed Mater Res A. 97(2):135-144. doi: 10.1002/jbm.a.33036. Epub 2011 Mar 2.

[110] Tugulu S, Löwe K, Scharnweber D, Schlottig F (2010) Preparation of superhydrophilicmicrorough titanium implant surfaces by alkali treatment. J Mater Sci Mater Med. 21(10):2751-2763.

[111] Tsukimura N, Ueno T, Iwasa F, et al. (2011) Bone integration capability of alkali- and heat-treated nanobimorphic Ti-15Mo-5Zr-3Al. Acta Biomater. 7(12):4267-4277.

[112] Zhang EW, Wang YB, Shuai KG, et al. (2011) In vitro and in vivo evaluation of SLA titanium surfaces with further alkali or hydrogen peroxide and heat treatment. Biomed Mater. 6(2):025001.

[113] Gaggl A, Schultes G, Müller WD, Kärcher H (2000) Scanning electron microscopical analysis of laser-treated titanium implant surfaces--a comparative study. Biomaterials. 21(10):1067-1073.

[114] Park CY, Kim SG, Kim MD, Eom TG, Yoon JH, Ahn SG (2005) Surface properties of endosseous dental implants after NdYAG and CO2 laser treatment at various energies. J Oral Maxillofac Surg. 63(10):1522-1527.

[115] Rajesh P, Muraleedharan CV, Komath M, Varma H (2011) Laser surface modification of titanium substrate for pulsed laser deposition of highly adherent hydroxyapatite. J Mater Sci Mater Med. 22(7):1671-1679.

[116] Wang H, Liang C, Yang Y, Li C (2010) Bioactivities of a Ti surface ablated with a femtosecond laser through SBF. Biomed Mater. 5(5):054-115.

[117] Palmquist A, Johansson A, Suska F, Brånemark R, Thomsen P (2011) Acute Inflammatory Response to Laser-Induced Micro- and Nano-Sized Titanium Surface Features. Clin Implant Dent Relat Res. [Epub ahead of print].

[118] Furuhashi A, Ayukawa Y, Atsuta I, Okawachi H, Koyano K (2011) The difference of fibroblast behavior on titanium substrata with different surface characteristics. Odontology. [Epub ahead of print].

[119] Galli C, Macaluso GM, Elezi E, et al. (2011) The effects of Er:YAG laser treatment on titanium surface profile and osteoblastic cell activity: an in vitro study. J Periodontol. 82(8):1169-77.

[120] Tsang CS, Ng H, McMillan AS (2007) Antifungal susceptibility of Candida albicans biofilms on titanium discs with different surface roughness. Clin Oral Investig. 11(4):361-368.

[121] Cimenoglu H, Gunyuz M, Torun Kose G, Baydogan M, Ugurlu F, Sener C (2011) Micro-arc oxidation of Ti6Al4V and Ti6Al7Nb alloys for biomedical applications. Materials Characterization 62(3): 304-311.

[122] Ma C, Nagai A, Yamazaki Y, et al. (2012) Electrically polarized micro-arc oxidized TiO(2) coatings with enhanced surface hydrophilicity. Acta Biomater. 8(2):860-865.

2

Rehabilitation Technologies: Biomechatronics Point of View

Erhan Akdoğan and M. Hakan Demir
Yıldız Technical University
Turkey

1. Introduction

Rehabilitation aims to bring back the patient's physical, sensory, and mental capabilities that were lost due to injury, illness, and disease, and to support the patient to compensate for deficits that cannot be treated medically (http://www.ehendrick.org/healthy, June 2010). After the Spinal Cord Injury (SCI), stroke, muscle disorder, and surgical operation such as knee artroplasticy, patients need rehabilitation to recover their movement capability (mobilization) (Bradly et al., 2000; Inal, 2000; Metrailler et al., 2007; Okada et al., 2000; Reinkensmeyer, 2003 and http://www.manchesterneurophysio.co.uk, November 2010). The number of those who need rehabilitation is steadily increasing everyday. Parallel to this, equipment and techniques used in the field of rehabilitation are becoming more advanced and sophisticated.

On the other hand, mechatronics, an interdisciplinary science, is a combination of machinery, electric-electronics and computer sciences plays an important role in rehabilitation technologies. In particular mechatronics systems provide important benefits for movements that are related to physical exercises in rehabilitation process.

Biomechatronics is a sub-discipline of mechatronics. It is related to develop mechatronics systems which assist or restore to human body. A biomechatronic system has four units: Biosensors, Mechanical Sensors, Controller, and Actuator. Biosensors detect intentions of human using biological reactions coming from nervous or muscle system. The controller acts as a translator between biological and electronic systems, and also monitors the movements of the biomechatronic device. Mechanical sensors measure information about the biomechatronic device and relay to the biosensor or controller. The actuator is an artificial muscle (robot mechanism) that produces force or movement to aid or replace native human body function (http://www.ece.ncsu.edu/research/bee/biomd, ND). Typical usage area of biomechatonics is orthotics, prosthesis, exoskeletal and rehabilitation robots, neuroprosthesis. In this chapter, rehabilitation robots will be discussed in terms of bio-mechatronics systems.

Especially in the last ten years, the number of studies about robots in the rehabilitation area has increased due to developments in actuator, sensors, computer and signal processing technologies. Some important reasons for the utilization of robots in rehabilitation can be listed as follows (Krebs, 2006):

- Robots easily fulfill the requirements of cyclic movements in rehabilitation;
- Robots have better control over introduced forces;
- They can accurately reproduce required forces in repetitive exercises;
- Robots can be more precise regarding required therapy conditions.

Rehabilitation robots can be classified into four groups:

- To assist disabled people in special need with their daily life activities,
- To support mobility,
- To assist social rehabilitation (Cognitive robotics)
- To assist therapists performing repetitive exercises with their patients (therapeutic exercise robot).

In this chapter, rehabilitation robots are introduced and explained in terms of functionality, control methods and equipment technology. The theory and terminology is given in Section II, first devices and tools are explained in Section III, rehabilitation robots are explained in terms of mechatronics in Section IV, types of rehabilitation robots are described in Section V.

2. Terminology of rehabilitation

2.1 The exercise movements in physical therapy and rehabilitation

There are five basic movements in Physical Therapy and Rehabilitation (Griffith, 2000). They are:

- *Extension*: The act of straightening or extending a limb.
- *Flexion*: The act of bending a joint or limb in the body by the action of flexors.
- *Abduction*: Abducting of a limb from middle line of body
- *Adduction*: Adducting of a limb to middle line of body
- *Rotating:* Rotating of a limb

2.2 Exercise types in physical therapy and rehabilitation: (Kayhan, 1995)

Passive Exercise: This exercise is performed for the patient by another person (nurse or therapist) or by an exercises device (robotic device or CPM). It is usually applied to patients who do not have muscle strength.

Active Assistive Exercise: As the patient develops the ability to produce some active movement, active exercise begins. Assistance can be provided manually by a therapist, by counterbalancing with weights or by gravity. This exercise is helpful in increasing the strength of the patient.

Active Exercises: These are the purposeful voluntary motions that are performed by the person himself, without resistance and with or without the aid of gravity. Active exercises can be resistive such as isotonic, isometric and isokinetic. They are called *resistive exercises*, as well. **Isometric** exercise involves the application of constant resistance to the patient on the range of motion (limb angle does not change). **Isotonic exercise** is applied to strengthen patients who suffer muscle contractions. Unlike isometric exercise, a constant resistance force is applied to the patient for the duration of the movement. **Isokinetic exercises** refer to resistance exercise performed at a constant preset speed. The speed is kept constant by resisting accommodating to muscle effort (torque). Isokinetic exercise should be applied by a machine. There some isokinetic machines such as BIODEX, CYBEX, Kin COM etc.

Manual exercise: These kinds of exercises are applied by physiotherapist. Physiotherapist decides that which type of exercise should be applied to patient. At this point, expertise of PT has a significant importance.

3. Conventional tools and devices in rehabilitation

In physical therapy and rehabilitation, some tools such as dumbbell, elastic band, rope and some mechanism in order to perform range of motion and strength exercise have been used for long time. At the end of 1970's, a significant development occurred. First electromechanical device in rehabilitation named as continuous passive motions (CPM) was developed and used (Fig 1). CPM gained a new momentum about usage of new technologies in rehabilitation. It was able to perform repetitive exercise motion on desired velocity and duration.

Fig. 1. Continuous Passive Motion Device (http://www.arthroscopy.com/sp06001.htm, October 2011)

The first clinical results about CPM's treatment effect were released by Salter and Simmonds in 1980. (Salter & Simmonds, 1980)

Then the computer controlled isokinetic machines which can also perform passive, isotonic and isometric exercises were developed. These machines were able to record exercise results. They are very important in order to follow rehabilitation period. The best known isokinetic machines are BIODEX (Fig.2), CYBEX and KinCOM. In this machine, a motor is used to generate resistance to movement. In previous models generated torque value was obtained via calculations. However modern isokinetic machines have load cell or force sensor to this regard. Today, they are being used in many medical centers.

At the middle of 1980's, some studies which were related to rehabilitation robotic were started to realize. For instance, Khalili & Zomlefer developed a robotic system for rehabilitation of joints and estimation of body segment parameters (Khalili & Zomlefer, 1988). Howell et al. used a robotic manipulator in order to rehabilitate the children (Howell et al., 1989). Kristy and et al. developed a smart exercise system which allows that the patient can perform some rehabilitative tasks with robot and computer. (Kristy et al., 1989)

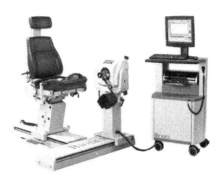

Fig. 2. BIODEX Isokinetic Machine (http://www.biodex.com, October 2011)

They were the first research studies on rehabilitation robotic. Detailed information about them is given in following sections.

4. Rehabilitation robotics in terms of mechatronics

4.1 What is mechatronics?

The definition of mechatronics has evolved since the original definition by the Yaskawa Electric Company. In trademark application documents, Yaskawa defined mechatronics in this way (Kyura & Oho, 1996; Mori, 1969)

> *"The word, mechatronics, is composed of "mecha" from mechanism and the "tronics" from electronics. In other words, technologies and developed products will be incorporating electronics more and more into mechanisms, intimately and organically, and making it impossible to tell where one ends and the other begins."*

At recent year another definition suggested by W. Bolton (Bolton, 1999);

> *"A mechatronic system is not just a marriage of electrical and mechanical systems and is more than just a control system; it is a complete integration of all of them."*

The interaction of the other areas and place of the Mechatronics system design are shown below clearly. (Craig, 2010)

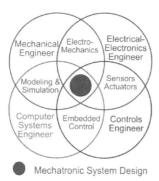

Fig. 3. Mechatronics System Design

As it is shown at the figure, Mechatronics is located at the middle of the clusters. Mechatronics system design involves Mechanical Eng., Electric-Electronic Eng., Control Eng. and Computer Eng. The study of Mechatronics systems can be classified as follows. (Bishop, 2002)

- Physical Systems Modeling
- Sensors and Actuators
- Signals and Systems
- Computers and Logic Systems
- Software and Data Acquisition

Robotics, motion control and conventional mechatronics, intelligent mechatronics, Human supported/based mechatronics, IT based mechatronics, micro and nano mechatronics and biomechatronics are application areas of mechatronics. (Akdoğan, 2004)

Biomechatronics and biomedical applications of mechatronics are emphasized in this chapter. Bio-industry requires mechatronics systems which have high quality and high precision. At Bio-Industry mechatronic systems are used for surgery, rehabilitation and prosthesis applications. Some biomechatronics systems are designed for these purposes are shown below.

a. Da Vinci Surgery Robot (http://www.roswellpark.org/robotics/about/da-vinci-surgical-system, December 2011)

b. Bionic Hand (http://www.BeBionic.com, December 2011)

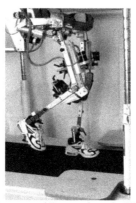

c. Locomat (Bernard et al., 2005)

Fig. 4. Biomechatronics Systems

Biomechatronics refers to a subset of mechatronics where aspects of the disciplines of biology, mechanics, electronics and computing are involved in the design and engineering of complex systems to mimic biological systems. Other definition of biomechatronics is suggested as follows:

"Biomechatronics is an applied interdisciplinary science that aims to integrate mechanical elements, electronics and parts of biological organisms. Biomechatronics includes the aspects of biology, mechanics, and electronics. It also encompasses the fields of robotics and neuroscience. (http:// www.wikipedia.com/biomechatronics, 2011)

Another definition of biomechatronics is suggested by H. Witte (Witte et al., 2005)

"Biomechatronics is the development and optimization of mechatronic systems using biological and medical knowledge. This strategy may be exemplified by bionically inspired robotics: identification of biological principles and their transfer into technical solutions extends the engineer's toolbox.

4.2 Actuators

In many mechatronics systems, movements or reactions to a situation are available. These movements and reactions are provided by system elements which produce force and torque. The displacement and acceleration are provided in the moving parts by these produced torque and force. Actuators are the elements in the systems that produce these torque and force. Electrical Motors, Hydraulic-Pneumatic actuators and their applications in rehabilitation robotics are mentioned in this chapter.

4.2.1 Electrical motors

Electromechanical actuators convert electrical energy into mechanical energy. Electromechanical actuators are divided into varieties between their energy converter mechanism such as electromechanical, electromagnetic, electrostatic and piezoelectric. The differences in electric motors are mainly in the rotor design and the method of generating the magnetic field. Some common terminologies for electric motors are: (George & Chiu, 2002)

The components of electric motors;

- Stator: It generates the appropriate stator magnetic field. It can be made of permanent magnets or copper windings. Magnetic field is created with these magnets and copper windings in the motor. Stator can be found inside or outside of the motor.
- Rotor: is the rotating part of the motor. Depending on the construction, it can be a permanent magnet or a ferromagnetic core with coil windings (armature) to provide the appropriate armature field to interact with the stator field to create the torque.
- Armature: is the rotor winding that carries current and induces a rotor magnetic field.
- Air Gap: It is the small gap between the rotor and the stator, where the two magnetic fields interact and generate the output torque.
- Brush is the part of a DC motor through which the current is supplied to the armature (rotor). For synchronous AC motors, this is done by slip rings.
- Commutator is the part of the DC motor rotor that is in contact with the brushes and is used for controlling the armature current direction.

Electrical motors are classified two major types as DC Motors and AC Motors. Because of the industrial requirements, there are many types of motors outside of three two major types such as stepper motors, linear motors and servo motors. These five major electric motor types are used in rehabilitation systems generally.

4.2.1.1 DC motors

DC motor is an electric motor that it converts direct current electrical energy into mechanical energy. Rotary movements of dc motor are smooth, precise and powerful. Speed of motor is in directly proportional to applied voltage. Output torque of the motor is directly proportional to armature's windings current strength. Structure of the dc motor is shown below.

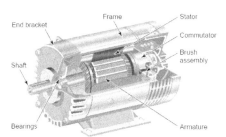

Fig. 5. Major Components of DC Motor (Petruzella, 2010)

As it is shown in the figure, dc motor contains electrical and mechanical components such as stator, armature, commutator, brushes, body and shaft.

DC motors are classified as brushed and brushless based on the presence of brushes in their structure. The induced current to the motor is transferred to the armature in two ways such as using brush and commutator or using electrical circuits. Brushed DC motor is a dc motor that it realizes commutation process with brushes and collectors. Brushed dc motors are classified as permanent magnet dc motors and electromagnet dc motors based on the source of the using magnetic field in their body. Permanent magnet dc motor use permanent magnets to produce stator's major magnetic field flux and use electro-magnets to produce armature's flux. Movement of the armature's magnetic field is obtained by the switching currents which flow in the armature's windings. Brushless DC motor hasn't collectors and brushes. Rotary parts of the motor consist of permanent magnets and stationary parts consist of miniature windings. The currents flow on the windings is controlled by the optical or magnetic sensors. This type of motor produces high torque.

DC motors are classified three types such as serial dc motors, shunt dc motors and compound dc motors based on the methods of excitation.

DC Motor is often use in rehabilitation robotics. It is used in applications not requiring high precision. For example LOKOMAT is a rehabilitation robot that it has four rotary joints. Stretching and expansion movements of the hip and knee are generated with these rotary joints. These movements of rotary joints are generated by dc motors. (Bernard et al., 2005)

MIME is a rehabilitation robot which uses PUMA 560 robot manipulator system to treat upper limbs of the patients. MIME use dc motors and gearboxes for positioning and orientation of end effector (gripper) of the robot (Lum et al., 1995, 1997).

4.2.1.2 AC motors

Ac motor is an electrical motor that converts alternative current electrical energy into mechanical energy. It is classified two major types such as induction motor and synchronous motor. Basic principle of these motors is that a mass is manufactured with magnetic plate is draft by rotary electromagnetic field.

Induction motor is preferred much than DC motor because of its cheapness and less maintenance requirement. Induction motor has stator, armature, body, supports and propeller. Armature windings are too much and stator's magnetic field induces current to armature. Armature doesn't connect any other energy supply. Rotation speed of the induction motor is changed slightly by load and rotation speed is regulated step by step. Speed of armature is always less than speed of stator's magnetic field because of the load on the motor. Induction motors are divided two types accoording to count of phase such as single-phase and three-phase induction motors and are divided two types according to their structure such as squirrel-cage induction motors and wound-rotor induction motors.

Synchronous motor is an alternative current electrical machine which rotates constant speed in proportion to frequency and pole numbers. Magnetic field is generated by current which is supplied in armature windings. This magnetic field is constant. Thus rotary field speed and speed of armature are same for this motor. This motor is used for applications requiring high power. As a result of synchronous rotation of the motor, sliding doesn't occur. Synchronous motors are classified two types based on the structure of armature such as salient pole synchronous motors and roundly pole synchronous motors.

4.2.1.3 Servo motors

Servo motors are used very effectively in velocity and displacement control of electro mechanical applications. Servo motors (Fig 6.b) have continuous motion and used feedback signals for velocity and displacement control. In servo motors the position of rotor is feedback to the controller (Fig 6.a) so the position error is minimized. (Petruzella, 2010)

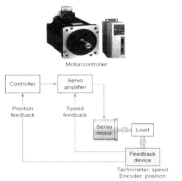

a. Closed loop systems of Servo motors   b. Servo Motor

Fig. 6. Closed loop systems of servo motors and servo motor (Petruzella, 2010)

In close loop servo-mechanism system as illustrated above position and velocity of motor measured with encoder and tachometer respectively and feedback to the controller and servo amplifier. Servo motors, unlike to the step motors, can be driven both AC and DC. Generally DC servo motors, AC servo motors and Brushless AC or DC motors are used in applications. DC servo motor design is similar to conventional DC motor design. A strong magnetic field is generated on DC servo motors to obtain high torque and excessive loadability. General properties of servo motors are; they need less energy according to conventional motors, they have small motor diameters with respect to standard DC motors, generated torque is proportional to rotor diameter and they have small inertia moments. DC servo motors are generally used in robotic applications that need high power. AC servomotors are of the two-phase squirrel-cage induction type motors. Main advantages of this type motors are, high safety, need less maintenance and not include commutators and disadvantages are; low efficiency, high heat dissipation and the requirement of AC square wave power source. In brushless DC servo motors rotor and stator is replaced. As a result of this brush and collector systems are cancelled and electro mechanical dissipations, which are based on friction and resistance between brush/collector systems, are disappearing. It is possible to obtain high torque from these motors. But the main disadvantage is a circuit must replace with collector, means that extra hardware requirement. Due to the precision of servo motor, they find spread place in robotic applications. Applications of servo motor in rehabilitation robots are explained in examples illustrated below.

Kikuchi et al., developed a rehabilitation system named Hybrid-PLEMO for it can be switched between active type and passive type and used with ER clutch/brake tool. They used AC servo motor and obtained two directional rotations with slow motions. They distribute these rotations on two ER tool. ER tool 1 is controlled the clockwise rotations and ER tool 2 is controlled the anti-clockwise rotations. (Kukichi & Furusho, 2009)

4.2.1.4 Stepper motors

A stepper motor (Fig. 7) converts electronic pulse into proportionate mechanical movement. Stepping motors are attractive because they can be controlled directly by computers or microcontrollers. Motor's shaft rotation quantity is in direct proportional to pulse number and speed of the motor is in direct proportional to frequency of this pulse. Rotate angle of the motor per each pulse determine the resolution of the motor. Stepper motor is used for open loop control systems generally. Produced movement for each pulse is precise and repeatable. Thus stepper motor is used for load positioning systems effectively. Step motor produces less power than 1 hp.

Fig. 7. Stepper Motor (Petruzella, 2010)

Advantages of stepper motor are that it doesn't need feedback, any positioning error doesn't occur during the motor movement and it doesn't need maintenance because of its simple structure. Disadvantages of this motor are that its movement is not continuous, it is less powerful than other motors and it may cause positioning errors in open loop control of friction based load applications. Stepper motors are classified three types such as variable-reluctance stepper motors, permanent magnet stepper motors and hybrid stepper motors.

4.2.1.5 Linear motors

Linear actuator is an actuator that creates motion in straight line, as contrasted with circular motion of a conventional electric motor (Boldea & Nasar, 1997). Linear actuators are classified types such as mechanical linear actuators, hydraulic and pneumatic linear actuators, piezoelectric linear actuators, electro-mechanics linear actuators and telescoping linear actuators.

In industry linear motor is used for applications requiring long distance and high sensitivity. Linear motor converts electrical energy into linear mechanical movement. Linear motor has had its stator and rotor "unrolled" so that instead of producing a torque (rotation) it produces a linear force along its length. Many designs have been put forward for linear motors, falling into two major categories, low-acceleration and high-acceleration linear motors. Low-acceleration linear motors are suitable for maglev trains and other ground-based transportation applications. High-acceleration linear motors are normally rather short, and are designed to accelerate an object to a very high speed, for example see the railgun. The main advantages of any linear motor is that it totally eliminates the need cost and limitations of mechanical rotation-to-translation mechanism. Thus the complexity of the mechanical system is drastically reduced. (http://www.dynamicdevices.com, December 2011). Its other advantages are that it works high speeds and high accurate, it response faster than mechanical transmission and there is no mechanical linkage in linear motor thus the stiffness of the motor increase. Disadvantages of this motor are that its performance is are affected by the temperature directly, it need an interface unit to connect the machine control component of the system.

Zhang use a linear motor in his rehabilitation robot in medical practice is to recover the function of the motor system of the injured limbs and trunk. His motor system issues break into two distinct categories: one is related to biomechanical/biophysical applications and the other to motor learning. Motion-speed and motion-range of a limb is limited by injury, burns, or postoperative conditions in that skin, ligaments, and muscles are inelastic from scar tissue. There are two modules in the Continuous Passive Motion Motion (CPM) mechanism for index fingers: biomimetic finger module and biomimetic muscle module. The two modules are linked by biomimetic muscle. The CPM mechanism can be driven from long-distance with the biomimetic muscles. Thus the biomimetic muscle consists of two pulleys, four spring bushings, two springs, a cord and a linear motor. The transmission distance between the linear motor and the CPM mechanism is adjustable by regulating the length of springs and cords. (Fuxiang, 2007)

4.2.2 Hydraulic and pneumatic actuators

4.2.2.1 Hydraulic actuators

Hydraulic systems (Fig. 8) are designed for moving heavy loads with high pressurized fluid in pipes and pistons which controlled with mechanical or electro mechanical valves. The figure shown above illustrates a hydraulic system and its' the components. A hydraulic system consist a tank, hydraulic pump, pressure regulator, valves and pistons. Tank is the storage member of the hydraulic fluid. Hydraulic pumps are generally drived by electric motors. At first, pressurized fluid is adjusted in pressure regulator for the hydraulic system. Regularized fluid is canalized by valves and actuated the hydraulic piston by this way linear motion is obtained. Most common hydraulic actuators are pistons and hydraulic motors. There are two types of piston used, one is single acting and the other is double acting, and hydraulic motors can be classified as gear motor, gerotor motor, and radial piston motor. Hydraulic actuators have great advantages like; have large lifting capacity, can used with servo control, have quick reaction capability, have self-cooling, have smooth motions at low speeds. But hydraulic systems are expensive, not appropriate for circular motion at high speeds and it is difficult diminish the extent of the system.

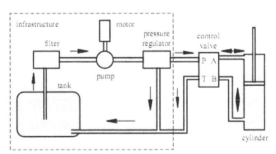

Fig. 8. Major Components of Hydraulic System

4.2.2.2 Pneumatics actuators

Pneumatic systems are moved with high pressurized gas in pipes and pistons which controlled with mechanical or electro mechanical valves. Pneumatic system consists of compressor, air conditioner unit and reservoir for gas storage, valves and piston. Working principle for pneumatic systems is similar to the hydraulic systems. One significant difference is pneumatic systems used compressible gas. Most common pneumatic actuators are pistons.

General advantages of pneumatic actuators are cheap, quick, and clean and safety on the other hand, desensitisation because of the gas compression, noise pollution because of the exhaust, gas leaks, needs extra drying and filtering and it is difficult to speed control. The disadvantage of pneumatic systems is that control of the pneumatic actuators is very difficult so their usage in rehabilitation robots is restricted. Some of the publications in this subject summarized below.

In their study Klute et al. use pneumatic muscles for bottom knee prosthesis. It aims to bring into connection between force, length, speed and activation with the myo-tendinous actuator and in this actuator flexible pneumatic muscles, hydraulic dampers and springs are used. (Klute et al., 1999)

Girone at al. design a Stewart platform-type haptic interface to use in rehabilitation systems. The system supplies six-DOF resistive forces in response to virtual reaility-based exercises running on a host PC. The used Stewart platform has double acting pneumatic cylinders, linear potentiometers as position sensors and six-DOF force sensor. (Girone et al., 2001)

4.3 Sensors in rehabilitation

We can examine sensors in two classes: mechanical and biological sensor.

4.3.1 Mechanical sensors

Force sensor: A Force Sensor is defined as a transducer that converts an input mechanical force into an electrical output signal. Force Sensors are known as Force Transducers, as well. Force sensor can measure forces in three axes. The device that is measure in single axis is called as "load cell". Force sensors are usually used for measuring applied or reaction force. Prostheses, therapeutic exercise robots and exoskeletal robot manipulator are common application area of force sensors in rehabilitation.

Position sensors: A position sensor is a device which allows position measurement. Position sensors can be either linear or angular. It is also known as potentiometer, rotary encoder or displacements sensor. Generally, it is used in mechanical systems for position feedback.

Goniometers: Goniometers are used to measure joint angle in rehabilitation. First and still used goniometers are mechanical. However, new types of electronic goniometers have been used in medical centres.

Hand Dynamometers: Hand Dynamometer is used to measure gross isometric power grip force.

Eye tracker: Eye tracker is a device that measures eye positions and eye movement. Eye trackers are especially used by disabled people to control a wheelchair.

4.3.2 Biological sensors

EMG system and sensors: The electromyogram (EMG) is an electrical response of the contracting muscle. Since their amplitude is quite small EMG signals are complicated. This is why EMG signals must be processed through signal analysis methods. In rehabilitation, EMG signals are used for two purposes, to evaluate patient muscle performance and to control a mechanical system such as a robot or prostheses.

EEG system and sensors: The electroencephalogram (EEG) is electrical activity of the brain. EEG recordings are obtained by placing electrodes of high conductivity in different locations of the head. Measures of the electric potentials are recorded between a pairs of active electrodes (bipolar recordings) or with respect to a supposed passive electrode called reference (monopolar recordings) as in EMG (Quiroga, 1998). Evaluating of EEG is harder than EEG. At the present time, many research studies have been done about that.

5. Rehabilitation robots

5.1 Assistive rehabilitation robotics

Robotic systems supporting daily tasks of patient such as holding, lifting, moving, free mobility are called as *assistive robotics*. Assistive robotics systems are developed in order to support disabled people to continue their life without any help from others.

There are some problems in the course of producing such technologies:

- Their high costs
- customized design requirements
- Robotic system can need many sensors according to disability level. This situation requires developing and using of high-performance computers, complex control methods and softwares.
- If patient cognitive capacity is not enough to use robotic system, patient could have difficulties in operating these systems. In this situation, safety problems can occur.

Target group of assistive robotics generally consist of people whose manipulation skills damaged, muscular dystrophy, cerebral palsy children, spinal cord injury. In particular, this robotic area that is related to daily life and individual is drawn attention with increasing population. On the other hand, its market share gains momentum day by day.

Mechanism and user interface design is very important in assisitve robotic systems. Interface is called as human-machine interface, human-computer interface and human-robot interface in the literature. Challenges of design of mechanism and user interface are as follows:

- Operator of system is not a specialist or engineer.
- Operator has some physical disables. Therefore, it needs a design that can enable to communicate between patient and system.
- Profiles and level of disabled is different. That is why; it needs to develop designs which can respond to needs of a wide-spectrum of disabled people and individuals.

It is required easy-to-use interfaces which provide interaction with system and useful mechanism in order to solve these problems. (Hammel, 1995)

The user interfaces are supported with several adaptive hardwares and sofwares such as speech recognition systems, track balls, special keyboard, eye-trackers- motion sensors etc.

Briefly a powerful user interface has features as summarized below:

- Individual
- Functional

- Adaptation capability
- Easy-to-use and understandable
- Funny

Assistive robotics can be detected in three groups: manipulation aid, mobility aid and social aid.

5.1.1 Manipulation aid

There are many studies have been done about manipulation aid robotic. Some of them are given follows:

ProVAR (Professional Vocational Assistant Robot):

The ProVAR robotic system is an assistive robot for individuals with a severe physical disability. It allows that a patient can control the system through a direct manipulation The ProVAR system has a PUMA-260 manipulator mounted on an overhead (see Fig 9). An optical emitter / detector, a force sensor and proximity sensors are integrated to the system in order to control the system and improve safety. Patient controls the system through a computer which has Windows-NT operating system. Computer also communicates with office devices such as telephone, fax and internet devices. (Van der Loss et al., 1999)

Fig. 9. ProVAR

AFMASTER: This robot was developed as a work station and works in 2m*3m area. The robot services goods which are in its office via user commands. AFMASTER allows severely disabled people to get back to work by automatic manipulation facilities. (Gelin et al., 2001)

DeVAR III – ADL:

DeVAR system includes a PUMA-260 industrial robotic arm, a computer and a wheelchair. End effector of robot manipulator is Otto Bock Greifer prosthetic hand. This hand can realize fingertip, cylindrical, and hook grasps. User uses the system via voice commands. The robot arm is surrounded by daily living equipment such as a microwave oven, a refrigerator, a tool holder for an electric shaver, a spoon, an electric toothbrush, pump toothpaste, adapted wash/dry cloths, and a mouth stick. Voice commands operate an X-10 environmental controller which supplies power to the robotic workstation, computers, lights, radio, and other appliances. (Hammel et al., 1987)

HANDY-1:

The Handy 1 is a rehabilitation robot designed to provide people with severe disability to gain/regain independence in important daily living activities such as: eating, drinking, washing, shaving, teeth cleaning and applying make-up. It has a robotic manipulator that has three degree of freedom. (Topping, 2002)

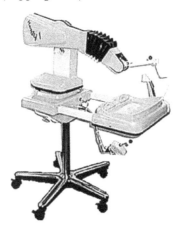

Fig. 10. Handy 1

5.1.2 Mobility aid

Mobility of disabled people in their houses and external environment is a very important phenomenon. To this regard, safe and reliable mobile robotic assistive devices are needed. These types of robots must be intelligent control structure. A state of art robotic device designed for mobility is able to:

- perform automated navigation tasks/behaviors through sensors,
- provide to access to users who may lack fine motor control,
- realize gesture recognition,
- recognize voice command
- do vision-based interaction,
- have ability to drive on all terrains such as stairs, curbs
- be adapted to intelligent homes, buildings communicate with smart wheelchairs
- Extend to other assistive devices (e.g. scooters)

To achieve these specifications sensors and software must be very powerful and control method of wheelchair must be flexible and effective. On the other hand, user interface between wheelchair and patient must be user-friendly.

Different input commands can be used in order to control a wheelchair. Traditional methods are joysticks, pneumatic switches and so on. Recently, hand voice recognition, eye tacking and machine vision techniques have been started to use on smart wheelchairs.

A wheelchair needs clear and correct information to perceive its surrounding. Therefore, sensors have a key role to control a wheelchair. Sonar sensors, infrared range finder, laser range finders, laser strips and cameras are being frequently used. Since their size and price sonar and infrared sensors are often used in most applications. However, they are not well suited to identify drop-offs such as stairs, cubs etc. That is why, laser rage finders are used. More accurate obstacle ad drop-off detection can be realized using them. On the other hand this type of sensors is very expensive and large. Mounting of them to a wheelchair is not easy. Laser striper consists of a laser emitter and a-charge coupled device camera. The laser striper cheaper than laser range finder but it can read false because of environment conditions. That is why, it has not been preferred. Briefly, a sensor must be cheap, accurate, small, lightweight, and must be not affected from environmental conditions such as temperature, lighting. On the other hand it has low energy consumption and can easily be mounted.

Researchers have developed different type of control methods and software. Some of them used artificial intelligence techniques whereas others multilayered control techniques including sensor - reactive behaviors.

Smart wheelchairs have several types of operating modes. a) Autonomous navigation with obstacle avoidance mode: wheelchair travels from its current positions to given positions b) Wall following mode: Operator controls wheelchair. Wheelchair avoids or stops in front of obstacle c) Door passage mode: Wheelchair passes from door with traverse movement d) Docking mode: Wheelchair enables close approach to object. e) Reverse trajectory: Wheelchair returns it's starting positions using recorded travel trajectory f) Target tracking mode: Wheelchair can follow and navigate an object g) Line following mode: Wheelchair can follow a line that was marked in environment. (Simpson, 2005)

There are some challenges faced by robotic device designed for mobility. Human intention and adapting it for a robotic device is not so easy. On the other hand, if users of device have a motor control problem this process will be harder. In terms of technology, designing of flexible input devices has a curious problem and cost of individual-oriented design can be high. Some well-designed mobility aid robotic device given is as follows:

PALMA: (assistive platform for alternative mobility)

PALMA is a robotic device to assist the mobility of children affected by CP (Fig 11). The PALMA has an open, safety and flexible structure; it can be adapted to different pathologies and varied motor dexterity. It allows to user a safe navigation using simple obstacle avoidance strategies. Especially, ultrasonic sensors were used for detection. In order to control the detection unit a PIC 16C73 microcontroller was used. Children can interact with environment via PALMA. (Ceres et al., 2005)

PAMM: (Personal Aid for Mobility and Monitoring)

PAMM is designed to assist the elderly people living independently (Fig. 12). It provides physical support and guidance, and it can monitor the user's basic vital signs. The PAMM has a force-torque sensor mounted under the user's handle to sense the user's intent. The system is controlled by an admittance based controller. It can communicate with a computer via wireless network in order to receive up- dated planning information and to provide information on the health and location of the user. (Dubowsky et al., 2000)

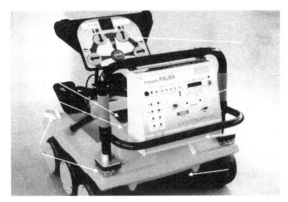

Fig. 11. PALMA (Ceres et al., 2005)

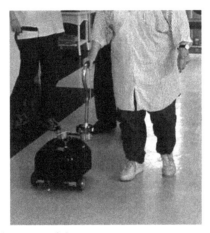

Fig. 12. PAMM (http://robots.mit.edu)

MAID (Mobility Aid for Elderly and Disabled People)

MAID is a smart wheel chair (Fig. 13). It is based on a commercial wheelchair that has been equipped with an intelligent control and navigation system. It allows using narrow and wide area. MAID can detect the environment objects as well as to identify which object is stationary and which is moving. Thus it can easily be used in crowded area. (Prassler et al., 2001)

Hong-Kong Advanced Robotic Lab. Smart Wheelchair:

A smart wheel chair was developed by the Chinese University of Hong-Kong Advanced Robotic Lab. This wheelchair has a human friendly interface for high-level control, and uses human skill models to significantly enhance the usability and functionality of traditional electric wheelchairs. To this regard a neural network structure was developed to map sensor readings to control actions to play back taught routes. (Hon Nin Chow & Yangsheng, 2006)

Fig. 13. MAID

Fig. 14. Smart wheelchair

5.1.3 Exoskeletal robot manipulators: (ExRM)

Exoskelelatal robot manipulators are used to provide or assist human motions. ExRM has a mechanism structure showing compliance with human joints and limbs. It supports human motions via actuators positioned in robot joints. ExRMs have been developed for industrial or military purposes. However, recent years research studies for which supporting physically disabled people movements about Ex RM have gained momentum. An ExRM interacts with a person directly. Therefore, control via biological signal processing stands out in these systems. On the other hand, it needs control structure based on artificial intelligence.

First studies have been started in 1960s. First ExRM was developed by Hardiman. Hardiman project aimed to increase human power on aircraft carriers, in underwater construction and on missions in outer space (Mosher, 1967)

Today research studies about usage of ExRM in rehabilitation area are being realized. A useful ExRM has features which are given below:

- Compatible with user's biomechanical features
- Safe
- Effective motion transfer
- Low power consumption
- Easy chargeable battery
- High performance control method
- High performance signal processing
- Small size powerful actuators

ExRMs can be classified three groups, lower extremity, upper extremity and full body systems. In these systems, electrical motors are generally used as actuators. On the other hand, in some studies pneumatic and hydraulic systems are used, as well. Rehabilitation purposed ExRMs studies are presented follows.

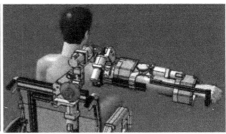

Fig. 15. (Moubarak et al., 2009)

Moubarak and *et al.* designed an ExRM for upper limb rehabilitation (Fig 15). The robot can perform repetitive exercise motions without physiotherapist. It has 4 DOF for shoulder, and elbow motions. Basic control method of system is force control. In particular, lightweight of mechanism was taken into account during design of the robotic mechanism. Servo motors were used as actuator. (Moubarak et al., 2009)

Fig. 16. ExRm of Tsai and et al. (Tsai et al. 2010)

Tsai et al. designed a 6DOF ExRM for passive and active exercises of upper limbs (Fig 16). Patient motion intension was sensed with EMG and force feedback combination. PID and impedance control methods were used to control of the system. Links are actuated with DC motors. (Tsai et al., 2010)

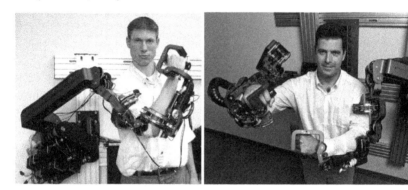

Fig. 17. Rosen and et al.'s 7DOF Manipulator EXO-7 (Perry, 2007; Yu & Rosen, 2010)

Yu and Rosen designed an anthropomorphic 7DOF Exrm (EXO-7) in order to assist shoulder, elbow and wrist motions (Fig 17). There is no backlash on the EXO-7. Its mechanical specifications are excellent. Therefore user can insert his arm easily to manipulator. It can be used for a therapeutic and diagnostics device for physiotherapy, an assistive (orthotic) device for human power amplifications, a haptic device in virtual reality simulation, and a master device for teleoperation. However, researchers used a new PID controller to guarantee the asymptotic stability in their last study. (Perry, 2007; Yu & Rosen, 2010)

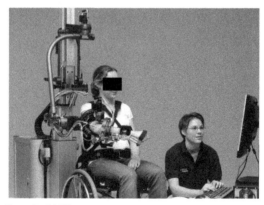

Fig. 18. ARMin (Mihelj et al., 2007)

ARMin is a 7DOF Exrm which has backdrivable DC motors and position-force sensors (Fig 18). It was used for training of patient for daily life activities. It uses admittance and impedance control methods which is best-known control methods in rehabilitation area. As all drives are backdrivable, a therapist can control the ARMin manually if it is necessary. (Mihelj et al., 2007)

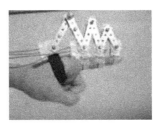

Fig. 19. (Yamura et al., 2009)

Yamaura and et *al*. developed a hand rehabilitation system. It consists of two units rehabilitation mechanism that moves fingers with actuators and a data glove that control the mechanism. The mechanism is wire-driven and all three joints of finger can be controlled with it (Fig 19) (Yamura et al., 2009)

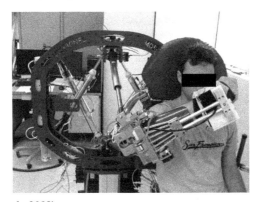

Fig. 20. Bones (Klein et al., 2008)

BONES (Biomimetic Orthosis for the Neurorehabilitation of the Elbow and Shoulder) is a 4DOF Exo-RM (Fig 20). It can perform arm internal/external rotation without any circular bearing element. This system uses pneumatic actuators. BONES is able to imitate a wide range of motion of the human arm. (Klein et al., 2008)

Fig. 21. SUEFUL-7 (http://www.me.saga-u.ac.jp/~kiguchi/research_en.html#research01, December 2011)

SUEFUL-7 (see Fig.21) is an EMG signal based 7DOF ExRM developed by Saga University. It can assist the motions of shoulder vertical and horizontal flexion/extension, shoulder internal/external rotation, elbow flexion/extension, forearm supination/pronation, wrist flexion/extension, and wrist radial/ulnar deviation. On the other hand researchers used a combination of impedance control method and muscle model oriented control in order to control of robot. Impedance parameters were adjusted through considering of upper limb posture and muscular contraction level. (Gopura et al., 2009)

5.1.4 Social assistive robotics

Social assistive robots help people by non-physical social interaction. The aim of social assistive robotics is to success purposes of rehabilitation and patient training. The robot interacts with stroke patient as therapeutic during daily life activities such as putting magazines to shelf.

5.2 Prostheses

Prosthesis is a mechanism which is developed in order to continue daily life activities of amputee people without others. Lower limb prostheses (Fig 22) are important in terms of providing mobility of person. However, people use their upper extremities much more for daily tasks such as lifting, pushing, gripping. That is why; one can find more about upper extremity prostheses.

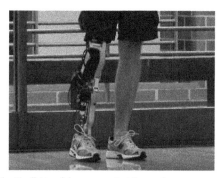

Fig. 22. Lower limb prosthetic (http://www.robaid.com)

Currently, myoelectric controlled prostheses are frequently used and reach studies which is related to this subject are being continued. The first study about prosthesis was realized by German physicist Reinhold Zeiter. Several studies about that were done in some countries such as Japan, USA and Rusia until 1960s. The myoelectric prosthesis called as MYOMOT which was developed by Otto Bock and Viennatone Comp. (Austrian manufacturer of hearing device) is regarded as the cornerstone in 1964 . (Inal, 2000)

There are some important challenges in producing prostheses:

- To control and coordinate multiple joints of a robotic limb
- To reach a high capability for a robot manipulator in terms of sufficient range of force, acceptable weight, portable power source and necessary sensors.

Useful, efficient and a high-technological prosthesis should have:

- Sufficient degree of freedom
- Resistant to environmental influences
- Sensitive artificial sensors,
- Well-designed cybernetic interface,
- Light and strong actuators,
- Long-life and forceful power source,

5.3 Therapeutic exercise robots

In general, a person with movement disabilities due to arm or leg problems needs to undergo periods of physiotherapy sessions spread over a long period of time. The sessions comprise a series of repeated and routine physical movements with the assistance and under the observation of a physiotherapist (PT). Transporting the patient to the medical center or calling a PT to where the patient is located are factors that further increase the cost of this process. The process of strengthening muscles to their normal values is costly and requires time and patience. Some important reasons for the utilization of robots in rehabilitation can be listed as follows; (Krebs, 2006)

- Robots easily fulfill the requirements of cyclic movements in rehabilitation;
- Robots have better control over introduced forces;
- They can accurately reproduce required forces in repetitive exercises;
- Robots can be more precise regarding required therapy conditions.

Because of these reasons, the number of studies about the usage of robots in rehabilitation has increased, especially in the last ten years.

During the rehabilitation process, patients sometimes move their extremities suddenly due to reflexes. Conventional machines such as CPMs do not respond in these kinds of situations and are hence not suitable for physical therapy. If a reflex causes a patient's leg to move while the machine is operating, an improper load results and can damage the patient's muscle or tendon tissue. (Sakaki et al., 1999) Because of this, there is a need for therapeutic exercise robots which can accomplish the rehabilitation of extremities based on the patient's complaints and real-time feedback during rehabilitation processes. Therapeutic exercise robots can be classified in two groups, upper limb and lower limb rehabilitation robots.

Therapeutic exercise robot systems for upper limbs:

Lee and others developed a robotic system for the rehabilitation of upper limbs of paralyzed patients using an expert system (Lee et al., 1990). This system combines the needed skills of therapists with a sensor-integrated orthosis and a real-time graphics system to ensure proper interaction and cooperation with the patient in order to achieve the goals of therapy. It can achieve passive exercises, motor-learning training and tone reduction.

Lum and *et al.* developed a prototype device called MIME (Mirror Image Motion Enabler) that implements passive and active assistive movements for upper extremities (Lum et al., 1995, 1997). This system uses two commercial robots (Fig 23).

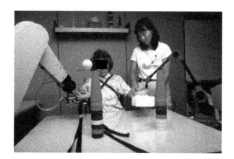

Fig. 23. MIME

Another system developed for rehabilitation of upper extremies is a robot manipulator with 5-DOF (degrees of freedom) called MULOS (https://www.asel.udel.edu, August 2008). It is used as an assistive device and to perform passive and resistive exercises. (See Fig 24)

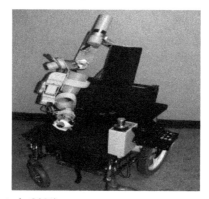

Fig. 24. MULOS (Johnson et al., 2001)

Krebs and others developed and have been clinically evaluating a robot-aided neurorehabilitation system called MIT-MANUS (Fig. 25) (Krebs et al., 1998, 2003). This device provides multiple degrees of freedom exercises of upper extremities for stroke patients. This study showed the effect of robot-aided rehabilitation.

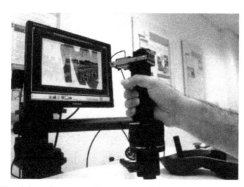

Fig. 25. MIT-Manus (http://coobox.wordpress.com)

Rao and others introduced another system using a Puma 240 robot for passive and active rehabilitation of upper extremities (Rao et al., 1999). In the passive mode, the robot moves the subject's arm through specified paths. In the active mode, a subject guides the robot along a predefined path, overcoming specified joint stiffness.

Richardson and et al. developed a 3-DOF pneumatic device (Fig.26) for the rehabilitation of upper extremities using PD control and impedance control methodologies (Richardson et al., 2003, 2005).

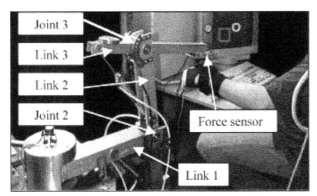

Fig. 26. 3-DOF pneumatic actuated device. (Richardson et al., 2003)

Another study for the rehabilitation of upper extremities is the REHAROB project, which uses industrial robots (Fig.27) (http://www.rehab.manuf.bme.hu, July 2008). A database is formed with the necessary force and position produced by the sensors placed on the patients during the rehabilitation process. Industrial robots then repeat the same procedure using this database.

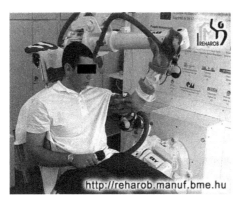

Fig. 27. REHAROB

Reinkensmeyer and others developed a 3-DOF system called ARM Guide (Assisted Rehabilitation and Measurement Guide) for the rehabilitation of upper extremities (Reinkensmeyer et al., 2000). The ARM Guide (Fig. 28) can diagnose and treat arm movement impairment following stroke and other brain injuries. As a diagnostic tool, it provides a basis for the evaluation of several key motor impairments, including abnormal tone, incoordination, and weakness. As a therapeutic tool, the device provides a means to implement and evaluate active assistive therapy for the arm.

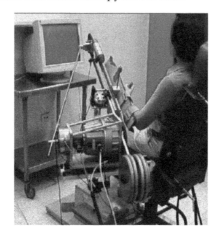

Fig. 28. ARM Guide

A 3-DOF system, called GENTLE/s (Fig.29), was developed for the rehabilitation of upper extremities using a haptic device and the virtual reality technique, controlled by the admittance control method (Luieiro et al. 2003).

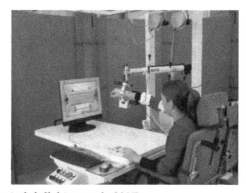

Fig. 29. GENTLE/s (Amirabdollahian et al., 2007)

Therapeutic exercise robot systems for lower limbs:

LOKOMAT (Bernhardt et al., 2005) and ALEX (Banala et al., 2007) were developed as gait rehabilitation robotic systems. The robotic systems for therapeutic exercises for lower limbs were developed to perform some repetitive, resistive, and assistive exercises which are performed by the PT or equipment.

Okada et al. employed an impedance control methodology in a 2-DOF robotic system (Fig.30), where the position and force data are received and recorded for the robotic system to imitate the corresponding motion (Okada et al., 2000)

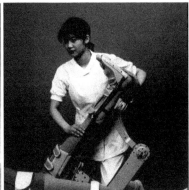

Fig. 30. TEM: Therapeutic Exercise Machine

Bradley et al. developed a 2-DOF autonomous system called NeXOS (Bradley et al., 2009) that is able to perform active assistive, passive and resistive exercises using pre-training visual position information. It can be used for knee and hip extension-flexion movements.

Moughamir et al. devised a training system called Multi-Iso (Moughamir et al., 2002). This system is also able to perform assistive, resistive and passive exercises like NeXOS. This 1-DOF system is used for knee limb and extension-flexion movements. Multi-Iso (Fig.31) uses classical force, position and speed control methods developed with fuzzy control techniques.

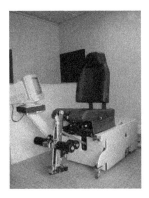

Fig. 31. Multi-ISO

Many of the developed therapeutic robotic systems are designed for only assistive or only passive and resistive exercises. Furthermore, few studies aim to model the PT's manual exercise and directly convey the PT's rehabilitation capability to the patient (http://www.rehab.manuf.bme.hu, July 2008; Okada et al., 2000; Akdogan et al., 2011)

Akdogan, Adli and Tacgin developed a 3DOF robot -Physiotherabot- for lower limb rehabilitation (Akdogan et al., 2009, 2011). Physiotherabot (Fig.32) can perform all active and resistive therapeutic exercises in addition to manual exercises of the PT for lower limbs. Meanly, it can work not only as a physiotherapy machine but also as a human (PT). On the other hand, Physiotherabot has 3-DOF and can perform flexion – the extension movement of the knee and hip, and the abduction-adduction movement of the hip.

Fig. 32. Physiotherabot

5.4 Cognitive robotics

Cognitive aid robots are very different from other types of therapy robots because they are actually agents of the therapist. They are not necessarily attached to the person who is getting physical therapy. For cognitive aid robots, also called communication aid robots, the goal is to provide a safe and friendly environment for the child to interact with other children or just to play in fact. The robots of this category are mostly focused on children who have cerebral palsy or children who have autism (Van der Loos, ND). There are many cognitive robots. Some of them given as follow:

Fig. 33. Cognitive robot CosmoBOT (http://www.anthrotronix.com, 2011)

Cosmo Bot (Fig.33) was developed to motivate children to participate more fully in therapy and education. CosmoBot was tested with children. Especially, it can be used for Downs Syndrome, cerebral palsy, muscular dystrophy, apraxia, neurodevelopmental disorders, and language developmental disorders. CosmoBot is a robot with nine degrees of freedom. It can be operated via computer-based software, and children can operate CosmoBot by using some input devices. (http://www.anthrotronix.com)

KASPAR (Kinesic and synchronisation in personal assistant robotics): KASPAR (Fig.34) is a child-sized humanoid robot developed by University of Hertfordshire. KASPAR is being used to study human-robot interaction that aims to build an open-source robot platform for cognitive development research. It was developed for robot assisted play for children with autism. KASPAR has 8 degrees of freedom in the head and neck and 6 in the arms and hands. It can open its mouth and smile. It can interact with children with its tactile sensors which is placed its skin surface. (http://kaspar.stca.herts.ac.uk)

Fig. 34. KASPAR

Another type of cognitive robots is pet robots. They communicate and interact with human. This interaction can be physically. They imitate animal reactions. The main purpose of pet robots is to support children and elderly people who need emotional and mental impairment using their imitation capability. They have several types of sensors such as visional, audial and tactile and have learning abilities for imitate emotions via interaction with human (Vander Loos; http:// www.aist.go.jp). There are many research studies about pet robots. In particular, **Paro Therapeutic robot** (Fig.35) is one of the best known. It allows the documented benefits of animal therapy to be administered to patients in environments such as hospitals and extended care facilities where live animals present treatment or logistical difficulties. It can reduce patient stress and their caregivers, stimulates interaction between patients and caregivers, and improves patient relaxation, motivation and socialization of patients. Paro has five kinds of sensors: tactile, light, audition, temperature, and posture sensors and 32-bit RISC processors. Moreover, it can understand light and dark. He feels being stroked and beaten by tactile sensor, or being held by the posture sensor. Paro can recognize the direction of voice and words such as its name, greetings, and praise with its audio sensor, as well (http://www.parorobots.com, 2011).

Fig. 35. Paro Therapeutic Robot (http://www.parorobots.com, 2011)

Keepon (Fig.36), is another example about pet robots developed by Michalowski, Mellon and Kozima. It has two video cameras (eyes) and a microphone (nose) as sensors, and four motors. It can remotely be control by a therapist in another room. Furthermore, it can be operated in an autonomous mode, bouncing along to music or sounds. Therefore, monitor, track, and record of the children's improvement can be record that it is very important to rehabilitation. (Michalowski et al., 2009)

Fig. 36. Keepon (Michalowski et al., 2009)

6. Future of rehabilitation robotics

Looking at the world's population developments in the field of rehabilitation robots will continue. Other factors which affect developments in this area are given follows:

- Hardware & Software
- Sensors
- Wireless communication
- "Always-on" computing

It is expected that the infrastructure of hardware and software will develop, high speed internet connection will spread and real time application performance will increase. On the other hand more stable, unaffected from environmental factors, high performance, small size and cheaper sensor will provide this development. Developing wireless communication systems, more portable devices and using them in home will have impact on the area of rehabilitation robot, as well. Patients are observed by rehabilitation team through "everywhere and everytime therapy" concept. Patients' treatment process is maintained through remote control applications and device without coming hospital.

So, patient, his/her relatives and health care personal will have more effective treatment period. Patients will be treated faster and they will contribute to their community as social and economic.

7. References

Akdogan, E. (2004). Yeni Bir Disiplin Olarak Mekatronik ve Türkiye'de Mekatronik Eğitimi, Proceedings of IV. Uluslararası Eğitim Teknolojileri Konferansı, Sakarya-Turkey, Kasım and 2004

Akdogan, E.; Tacgin, E. & Adli, M.A. (2009). *Knee Rehabilitation Using an Intelligent Robotic System, Journal of Intelligent Manufacturing, Springer Verlag.*, Vol:20, No:2, pp. 195-202

Akdogan, E. & Adli, M.A. (2011). The Design and Control of a Therapeutic Exercise Robot for Lower Limb Rehabilitation: Physiotherabot, *Mechatronics-Elsevier*, Vol: 21, No:3, pp. 509-522

Amirabdollahian, F.; Loureiro, R., Gradwell, E., Collin, C., Harwin, W. & Johnson, G. *Multivariate analysis of the Fugl-Meyer outcome measures assessing the effectiveness of GENTLE/S robot-mediated stroke therapy*, Journal of Neuroengineering and Rehabilitation (2007)

Banala, S.K.; Agrawal, S.K. & Scholz, J.P. Active Leg Exoskeleton (ALEX) for Gait Rehabilitation of Motor-Impaired Patients, *Proceedings of the IEEE 10th International Conference on Rehabilitation Robotics*, 2007, pp. 401 – 407

Bradley D.; Marquez C., Hawley M., Brownsell S., Enderby P. & Mawson S. (2009). NeXOS – The design, development, and evaluation of a rehabilitation system for the lower limbs, *Mechatronics*, 19, pp. 247-257

Bernhardt, M.; Frey, M., Colombo, G. & Riene, R. Hybrid force-position control yields cooperative behavior of the rehabilitation robot LOKOMAT, *Proceedings of the 9th International Conference on Rehabilitation Robotics*, 2005, pp. 536–539

Bishop, R. H. & Ramasubramanian, M. K. (2002). What is Mechatronics?, In: *The Mechatronics Handbook, R. H. Bishop, pp. (1-1,1-10)*, CRC Press, 0-8493-0066-5, USA

Boldea, I. & Nasar, S. A. (1997). *Linear electric actuators and generators*, Cambridge University Press, 0-521-48017-5, New York, USA

Bolton, W. (1999). *Mechatronics: Electrical Control Systems in Mechanical and Electrical Engineering* (2nd Ed.), Addison-Wesley Longman, 0-582-35705-5, Harlow, England

Ceres, R.; Pons, J.L., Calderón, L., Jiménez, A.R. & Azevedo, L. (2005). *A Robotic Vehicle for Disabled Children*, In: Engineering in Medicine and Biology Magazine, IEEE , Issue Date : Nov.-Dec.,Volume : 24 , Issue:6 pp. 55 - 63

Craig, K. (2010). Mechtronic System Design, In: *The Mathworks Virtual Academic Conference 2010, Data Access*, Available from:
<http://multimechatronics.com/images/uploads/home_Page/recent_Events/ma thWorks_2020/Mathworks%202010%20Academic%20Virtual%20Conference%20Pr esentation.pdf>

Dubowsky, S.; Genot, F., Godding, S., Kozono, H., Skwersky, A., Haoyong Y. & Shen Y. L. PAMM - a robotic aid to the elderly for mobility assistance and monitoring: a "helping-hand" for the elderly, *Proceedings of ICRA '00. IEEE International Conference*, 2000, vol.1, pp. 570 - 576

Fuxiang, Z. (2007) . An Embedded Control Platform of a Continuous Passive Motion Machine for Injured Fingers, In: *Rehabilitation Robotics*, Kommu, S. S., pp.(579-606), I-TEch Education and Publishing, 978-3-902613-01-1, Vienna, Austria

Gelin, R.; Lesigne, B., Busnel, M. & Michel, J.P. (2001). The first moves of the AFMASTER workstation, *Adv. Robotics*, 14, pp. 639–649

George, T. & Chiu, C. (2002). Electromechanical Actuators, In: *The Mechatronics Handbook, R. H. Bishop, pp. (20-1,20-33)*, CRC Press, 0-8493-0066-5, USA

Girone, M.; Burdea, G., Bouzit, M. & Popescu, V. (2001). A stewart paltform-based system for ankle telerehabilitation, *Autonomous Robots 10*, pp. 203-212

Gopura, R.A.R.C.; Kiguchi, K. & Yang L. SUEFUL-7: A 7DOF upper-limb exoskeleton robot with muscle-model-oriented EMG-based control, *Proceedings of International Conference on Intelligent Robots and Systems, IROS 2009*, 2009, pp. 1126 - 1131

Griffith, H.W. (2000). *Spor Sakatlıkları Rehberi*, Birol Basın Yayın Dağıtım Tic. A.Ş., İstanbul, Türkiye, pp. 451-452

Hammel, J. (1995). The role of assessment and evaluation in rehabilitation robotics research and development: Moving from concept to clinic to context, *IEEE Trans. Rehab. Eng.*, 3, pp. 56–61

Hammel, J. MA, OTR; Karyl Hall, EdD ; David Lees, MS ; Larry Leifer, PhD; Machiel Van der Loos, EdME ; Inder Perkash, MD ; Robert Crigler, (1987). Clinical evaluation of a desktop robotic assistant, *Journal of Rehabilitation Research and Development*, Vol . 26, No . 3, pp. 1-16

Hon Nin Chow, H. N. & Yangsheng Xu, Y. (2006). Learning Human Navigational Skill for Smart Wheelchair in a Static Cluttered Route, *IEEE Transactions On Industrial Electronics*, vol. 53, no. 4

Howell, R. D.; Hay, K. & Rakocy, L. Hardware and software considerations in the design of prototype educational robotic manipulator, *Proceedings of 12th RESNA Conf.*, Washington D.C., 1989, pp. 113–114

Inal S. (2000). *Kas hastalıklarında rehabilitasyon ve ortezler*, Cizge tanitim Ltd.Şir., Istanbul

Inal, S. (2000), *Üst ekstremite protezleri*, Cizge tanitim Ltd.Şir., Istanbul

Johnson, G.R.; Carus, D.A., Parrini, G., Scattareggia Marchese, S. & Vale, R. (2001). The design of a ve-degree-of-freedom powered orthosis for the upper limb, *Proc. Instn. Mech. Engrs.*, Vol 215, Part H.

Kayhan O. (1995). *Lectures and Seminars in Physical Medicine and Rehabilitation* (1st edition) Marmara Universitesi Yayınları, İstanbul, Türkiye

Khalili, D & Zomlefer M., (1988). Intelligent robotic system for rehabilitation of joints and estimation of body segment parameters, *IEEE Trans. Biomed. Eng.*, Vol. 35, no. 2, pp. 138-146

Kikuchi, T. & Furusho, J., (2009). "Hybrid-PLEMO", Rehabilitation system for upper limbs with Active/Passive Force Feedback mode, In: Recent Advances in Biomedical Engineering, Ganesh, G. N., pp. (361-376) , InTech, Retrieved from < http://www.intechopen.com/books/show/title/recent-advances-in-biomedical-engineering>

Klein, J.; Spencer, S.J., Allington, J., Minakata, K., Wolbrecht, E.T., Smith, R., Bobrow, J. E & Reinkensmeyer, D. J. Biomimetic Orthosis for the Neurorehabilitation of the Elbow and Shoulder (BONES), *Proceedings of IEEE/RAS-EMBS Int. Conf. on Biomedical Robotics and Biomechatronics*, 2008, pp. 535-541

Klute, G. K.; Czerniecki, J. M. & Hannaford, B. (1999). McKibben Artificial Muscles: Pneumatic Actuators with Biomechanical Intelligence, Proceedings of IEEE/ASME International Conference on Advanced Intelligent Mechatronics, Atlanta, September and 1999

Krebs, H.I.; Hogan, N., Aisen, M.L. & Volpe, B.T. (1998). Robot-aided neurorehabilitation, *IEEE Trans. Rehabil. Eng.*, 6, pp. 75-87

Krebs, H.I.; Palazzolo, J.J., Volpe, B.T. & Hogan, N. (2003). Rehabilitation robotics: performance based progressive robot assisted therapy, *Auton. Robotics*, 15, pp. 7-20

Krebs H.I. (2006). *An overview of rehabilitation robotic technologies. In: American Spinal*, Injury Association Symposium

Kristy, K.A.; Wu, S.J., Erlandson, R.F., deBear, P., Geer, D. & Dijkers, M. A Robotic arm - smart exercise system-: a rehabilitation therapy modality, *Proceedings of IEEE Engineering in Medicine 11th Annual International Conference,* (1989)

Kyura, N. & Oho, H. (1996). Mechatronics—an industrial perspective, *IEEE/ASME Transactions on Mechatronics*, Vol. 1, No. 1, pp. 10–15

Lee, S.; Agah, A. & Bekey, G. An intelligent rehabilitative orthotic system for cerebrovascular accident, *Proceeding of the IEEE International Conf. on Systems, Man and Cybernetics,* 1990, pp. 815-819

Loueiro, R.; Amirabdollahian, F., Topping, M., Driessen B. & Harwin W. (2003).Upper limb mediated stroke therapy – GENTLE/s approach, *Auton. Robotics,* 15, pp. 35-51

Lum, P.S.; Lehman, S., Steven, L. & Reinkensmeyer, D.J. (1995). The bimanual lifting rehabilitator: An adaptive machine for therapy of stroke patient, *IEEE Trans. Rehabil. Eng.*, 3, pp. 166-173

Lum, P.S.; Burgar, G. & Van Der Loos, M. The use of robotic device for post stroke movement therapy, *Proceeding of The International Conference on Rehabilitation Robotics*, 1997, pp. 79-82

Metrailler P.; Brodar R., Stauffer Y., Clavel R. & Frischknecht, R.. (2007). Cyberthosis: Rehabilitation robotics with controlled electrical muscle stimulation, *Rehabilitation Robotics, Itech Education and Publishing, Austria*

Mihelj, M.; Nef, T. & Riener, R. ARMin II-7 DOF Rehabilitation Robot: Mechanism and Kinematics, *Proceedings of IEEE Int. Conf. On Robotics and Automat.*, 2007, pp. 4120-4125

Mori, T. (1969). Mechatronics, *Yasakawa Internal Trademark Application Memo*, 21.131.01, July 12

Mosher, R. S. (1967). *Handyman to Hardiman*, Society of Automotive Engineers Publication, MS670088

Moubarak, S.; Pham, M. T., Pajdla, T. & Redarce, T. Design and Modeling of an Upper Extremity Exoskeleton, Proceedings of *Int. Congress of the IUPESM: Medical Physics and Biomedical Engineering*, 2009

Moughamir, S.; Zaytoon, J., Manamanni, N. & Afilal, L. (2002). A system approach for control development of lower limbs training machines, *Control Eng. Pract.*, 10, pp. 287-299

Okada S.; Sakaki T., Hirata R., Okajima Y., Uchida S. & Tomita Y. (2000). TEM, A therapeutic exercise machine for the lower extremities of spastic patient, *Adv.Robotics*, 14, pp.597-606

Prassler, E.; Scholz, J. & Fiorini, P. (2001). *A robotics wheelchair for crowded public environment*, IEEE Robotics & Automation Magazine, IEEE , Volume: 8 Issue:1 , pp. 38 - 45

Perry, J. C.; Rosen, J. & Burns, S. (2007). Upper-Limb Powered Exoskeleton Design, *IEEE/ASME Trans. on Mechatronics*, vol. 12, no. 4, pp. 408- 417

Petruzella , F.D. (2010). *Electric Motors Control Systems*, McGraw-Hill, 978-0-07-352182-4, New York, USA

Quiroga, R. (1998). *Quantitative analysis of EEG signals: Time-frequency methods and Chaos theory*, Ph.D. Thesis, Argentina.

Rao, R.; Agrawal, S.K. & Scholz, J.P. A robot test-bed for assistance and assessment in physical therapy, *Proceeding of the International Conference on Rehabilitation Robotics*, 1999, pp. 187-200

Reinkensmeyer, D.J.; Kahn. L.E., Averbuch. M., McKenna-Cole. A., Schmit. B.D. & Rymer, W.Z. (2000). Understanding and treating arm movement impairment after chronic brain injury: Progress with the ARM Guide, *Journal of Rehabil. Res. Dev.*, 37, pp. 653-662

Reinkensmeyer D.J. (2003). *Standard Handbook of Biomedical Engineering and Design Rehabilitators* (1st edition). McGraw-Hill, Columbus

Richardson, R.; Brown, M., Bhakta, M. & Levesley M.C. (2003). Design and control of a three degree of freedom pneumatic physiotherapy robot, *Robotica*, 21, pp. 589-604

Richardson, R.; Levesley, M.C., Brown M. & Walker P. (2005). Impedance control for a pneumatic robot-based around pole-placement, joint space controllers, *Control Eng. Pract.* 2005,13, pp. 291–303

Sakaki, T.; Okada, S., Okajima, Y., Tanaka, N., Kimura, A., Uchida, S., Taki, M., Tomita, Y. & Horiuchi, T. TEM: Therapeutic exercise machine for hip and knee joints of spastic patients, *Proceeding of the Sixth International Conference on Rehabilitation Robotics,* 1999, pp. 183-186

Salter R. & Simmonds B. W., (1980). The Biological Effect of Continuous Passive Motion on the Healing of Full Thickness Defects in Articular Cartilage: An Experimental Investigation in The Rabbit, *The Journal of Bone and Joint Surgery,* 62-A, pp. 1232-1251

Simpson, R. (2005). Smart wheelchairs: A literature review, *Journal of Rehabilitation Research and Development,* Vol. 42, No. 4, pp. 423-438

Tsai, B. C.; Wang, W. W., Hsu, L. C., Fu, L. C. & Lai, J. S. An Articulated Rehabilitation Robot for Upper Limb Physiotherapy and Training, *Proceedings of IEEE/RSJ Int. Conf. on Intelligent Robots and Systems,* 2010, pp. 1470-1475

TOPPING, M. (2002). An Overview of the Development of Handy 1, a Rehabilitation Robot to Assist the Severely Disabled, *Journal of Intelligent and Robotic Systems,* 34, pp. 253–263

Van der Loos, H.F.M.; Wagne, J.J., Smaby, N., Chang, K., Madrigal, O., Leife, L.J. & Khatib, O. ProVAR assistive robot system architecture, *Proceedings of the 1999 IEEE International Conference on Robotics & Automation Detroit,* Michigan, 1999, pp. 741-746

Van der Loos, H.F.M. Lecture Notes in Assistive Technologies

Witte, H.; Lutherdt S. & Schilling, C. (2005). Biomechatronics: how much biology does the engineer need?, *Proceedings of the 2004 IEEE International Conference on Control Applications,* 0-7803-8633-7, Taiwan, September and 2004.

Yamaura, H.; Matsushita, K., Kato, R. & Yokoi, H. Development of Hand Rehabilitation System for Paralysis Patient-Universal Design Using Wire-Driven Mechanism, *Proceedings of Int. Conf. of the IEEE EMBS,* 2009, pp. 7122-7125

Yu, W. & Rosen, J. (***) *A Novel Linear PID Controller for an Upper Limb Exoskeleton,* in Proc. of *IEEE Int. Conf. on Decision and Control,* pp. 3548-3553 (2010)

<http://www.manchesterneurophysio.co.uk>

<http://www.ece.ncsu.edu/research/bee/biomd>

<http://www.arthroscopy.com/sp06001.htm> (Access time: 11.10.2011)

<http://www.biodex.com> (Access time: 11.10.2011)

<http://robots.mit.edu>

<http://www.me.saga-u.ac.jp/~kiguchi/research_en.html#research01, Access time: December 2011>

<http://www.robaid.com/>

<www.asel.udel.edu>, Access Time: Aug. 13, 2008.

<http://coobox.wordpress.com>

<reharob.manuf.bme.hu>, Access Time: July. 30, 2008.

<http://www.anthrotronix.com>

<http://kaspar.stca.herts.ac.uk>

<http://www.aist.go.jp>

<http://www.parorobots.com>

<http://www.dynamicdevices.com>

<http:// www.wikipedia/biomechatronics> (Access Time: December 2011)

<http://www.BeBionic.com> (Access Time: December 2011)
<http://www.roswellpark.org/robotics/about/da-vinci-surgical-system> (Access Time: December 2011)

3

Biomedical Optics and Lasers

B. Cem Sener
Department of Oral and Maxillofacial Surgery
Faculty of Dentistry, Marmara University, Istanbul,
Turkey

1. Introduction

Electromagnetic energy is the source of life in our universe. Living organisms require energy to survive. Basic organisms provide this energy externally as a raw material (basic molecules like sugar) and convert it to energy (digestion of food). However some organisms, like green plants, can directly absorb radiation energy and store after a chemical conversion. Excessive energy is stored with a chemical reaction as adenine triphosphate (ATP) synthesis. Complex organisms digest more simple ones that contain readily synthetized ATP in an understanding of a food-chain. Under specific conditions complex organisms, partly, can absorb radiation energy and use it directly or store with ATP. Therefore, different bands located in certain parts of wide radiation energy spectrum can be used directly or indirectly to maintain our life. With this respect; interactions with radiation is mainly subjected within this context either with positive and negative ways, however both paths can be a matter of survival.

In that broad spectrum of radiation energy only a small part is visible (between 400 to 700 nm wavelength) for human beings (light). (Figure 1) However, mankind has converted or digitalized rest of spectrum to monitor and understand how it works for many purposes, like industry, navigation, communication, industry, research, military and medicine. Focus of this chapter is centered on a specific form of radiation, L.A.S.E.R. Amplified light generated by stimulated of emission can be defined as purified and well-disciplined form of radiation and therefore has a special importance in biomedical applications with its advanced features. Moreover; developments in optics, computer hardware and software, electronics and nanotechnology have widen the scope of LASER radiation in biomedical applications. This chapter aims to give a basic understanding of laser and its place as a solution for several issues struggled in clinical and laboratory environments of biomedical sciences. Innovative techniques centered on laser application are also given besides emphasized basic technical knowledge so that readers can develop their informative wallet to build up a vision for further steps both in bio- and engineering sciences. This would also have a collateral effect to prepare current reader as developers, researchers or users for the future.

2. History

Interaction of radiation energy with biologic systems is not a new concept and has been focus of human beings for millenniums. Evidences in ancient Greek writings for use of sun light as a therapy under "heliotherapy" term was first mentioned by Licht in 1983 [1]. During those ancient eras sun light (helios) or ancient Sun god was believed to have a blessing - healing (therapeutic) effect on living organisms. Healing effect of sun light has been a fundamental belief of esoteric teachings without any scientific basis. This belief continued during monotheistic religion for almost 400 years until rise of Holy Roman Empire. Even the suppression of ancient Sun worship by Christian orthodoxy; power of light has sustained in many teachings and illustrations of holy religions. Following development of observatory medical sciences in 18th and 19th centuries, benefits of sun light rediscovered for treatment of several diseases like scurvy, rickets, edema, rheumatic arthritis, dropsy, and depression caused by dark habitation environments during progress of coal industry. Increasing experimental sciences has proven that use of sun light has positive results in treatment of certain diseases and moreover development in optics and electromagnetic physics has brought evolvement of ultraviolet lamps and X-ray application in therapeutic and diagnostic medicine.

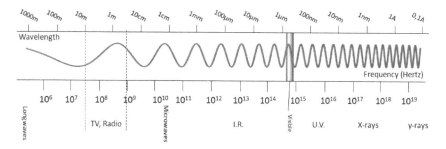

Fig. 1. Electromagnetic wave spectrum

Electromagnetic energy used in biotechnology has progressed with wider steps since "Zur Quantum Theorie der Strahlung" had been postulated by Albert Einstein in 1917. This explains the way how photoelectric amplification could emit a single frequency or stimulated emission which explains the way how laser works. In this direction; different groups worked on to improve more specialized and enhanced radiation energy source with augmented features next 40 years. In 1955 Gordon and coworkers described microwave amplification by stimulated radiation (MASER) using ammonia gas as the active medium, which can be termed as the ancestor of laser. Following, Schawlow and Townes (1958) applied this theory to light and named their invention as "optical maser". Although this development, Dr. Theodore Maiman (1960) pronounce "LASER" (Light Amplification by Stimulated Emission of Radiation) term for his device using ruby crystal to produce red laser beam at 694nm wavelength. Following the 1960s and rapid development in laser technology including the use of wide range of alternative lasing media; new wavelengths have been introduced to the market as summarized in table 1.

3. Laser physics – Components

Underlying physical principles relevant laser irradiation is basically same for all laser types. Fundamental components of a laser energy producing devices are illustrated in figure 2 and as follows:

a. Active lasing medium
b. Energy source
c. Optical (Resonating) cavity
d. Cooling system
e. Delivery system

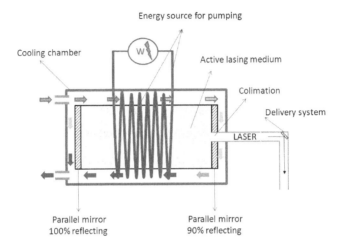

Fig. 2. Laser system illustration

3.1 Active lasing medium

Since the beginning of laser technology plenty of elements or mixtures are tested solely or in combination with another, regarding their potential to generate laser beam. Active lasing medium is a material, which can absorb energy given by the external source, to be excited through changes in the electrons of its atoms, molecules or ions, and finally emit this excessive energy as photons (Figure 3a,b,c).

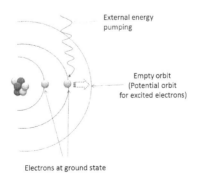

Fig. 3a. Stimulation (pumping phase). Excitation of the atom at ground state with external energy source.

Fig. 3b. Excited atom at higher energy status is ready to emit radiation while returning to its ground status.

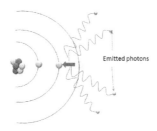

Fig. 3c. Photons are emitted while excited atom are returning to ground state following excitation phase.

Those releasing sequence (wavelength) is directly related with the type of lasing medium. Different lasing mediums can be used in combination to alter the predefined wavelength and/or carrier mediums that suspend the active lasing medium can also be excited and emit photons at different wavelength. This modification gives the possibility of generating two different laser wavelengths in one system, which can help to accelerate its effect on the target. Lasing mediums can be pure or in combination with a host material as a mixture or solution. Each medium has different photon emission nature that produces various wavelengths specific to the lasing medium. Features of laser beam is directly related with its wavelength (its place in the electromagnetic wave spectrum) and determined by its active lasing medium. Type of the laser is named with its wavelength on the radiation spectrum (like 810nm) or its active lasing medium (like diode laser). Active lasing mediums can be:

Gas

CO_2, argon, krypton and combinations like helium-neon and excimer lasers like argon-fluoride (ArF) 193 nm, krypton-fluoride (KrF) 248 nm, xenon-chloride (XeCl) 308 nm, xenon-fluoride (XeF) 353 nm are the most common types of gas lasers. Additionally xenon, nitrogen and carbon monoxide are used with different purposes. Additionally metal ions as a vapor can be used in combination with gas molecules to produce wide range of spectrum such as helium-silver (HeAg) 224 nm, helium-cadmium (HeCd) 441.563 nm, 325 nm, helium-mercury (HeHg) 567 nm, 615 nm, helium-selenium (HeSe) between red and UV, strontium vapor 430.5 nm, copper vapor (Cu) 510.6 nm, 578.2 nm, neon-copper (NeCu) 248 nm and gold (Ag) vapor 627 nm. Majority of the gas state laser mediums are pumped with electrical currency.

Liquid

Liquid lasing mediums have one dye material, which is active counterpart of lasing (mostly a dye) and the solvent carrier (commonly an organic chemical solvent). Wide range of chemicals can be used, like copper, chromium, dyes, metallic salts. The medium can be fluent or even more dens (gel form). Coumarin, rhodamine and fluorescein are used as dye materials as a solution in chemical solvents such as methanol, ethanol or ethylene glycol. Periodic dielectrical stimulation of liquid crystal can be used for modulation of wavelength and may help modifying wave forms and frequencies of the generated laser beam.

Solid

Solid state lasing mediums are commonly in glass or crystal form. Active lasing material carried by a host material. With this purpose most common host material is yttrium-aluminum-garnet (YAG) combination. Several active lasing materials like neodymium (Nd:YAG) 1.064 µm, erbium (Er:YAG) 2.94 µm, thulium (Tm:YAG) 2.0 µm, ytterbium (Yb:YAG) 1.03 µm, holmium (Ho:YAG) 2.1 µm have been combined with this YAG carrier. Also yttrium lithium fluoride (YLF) with neodymium (Nd:YLF-1.047 and 1.053 µm), sapphire (aluminum oxide), like Ti:sapphire-650-1100 nm, Cr:sapphire (ruby laser-694.3 nm), and silica glasses (Nd:Glass -1.062 µm, ytterbium$_2$O$_3$-1.03 µm, Promethium 147 doped phosphate glass (^{147}Pm^{+3}:Glass)- 933 nm, 1098 nm, erbium doped and erbium-ytterbium codoped glass-1.53-1.56 µm) are used. As shown, proper types and amounts of dopants can help to generate different types of wavelengths. Commonly flash lamps or other kind of lasers are used for pumping in beam generation.

Semiconductor

Semiconductors are solid state materials using semiconductors as gain media for lasing. Light is emitted at an interband transition (conduction band). Most frequently used semiconductor gain mediums are: GaAs (gallium arsenide), AlGaAs (aluminum gallium arsenide), GaP (gallium phosphide), InGaP (indium gallium phosphide), GaN (gallium nitride), InGaAs (indium gallium arsenide), GaInNAs (indium gallium arsenide nitride), InP (indium phosphide) and GaInP (gallium indium phosphide). Combination of different elements enables to modulate the band gap energy. Their rage in wavelength spectrum is much of the visible, near-infrared and mid-infrared spectral region. Semiconductor lasers are commonly laser diodes and pumped with an electrical current in a region where an n-doped and a p-doped semiconductor material meet. On the other hand, optically pumped semiconductor lasers are also used, where carriers are generated by absorbed pump light, and quantum cascade lasers, where intraband transitions are utilized. Most semiconductor lasers generate a continuous output. Pulses can be generated with switching on the pump source only for certain time intervals, long enough to fulfill lasering process (quasi-continuous-wave operation), which allows to gain significantly enhanced powers. Ultrashort pulses can also be generated with mode locking (optical technique) or gain switching (modulating pump power).

Biologic materials

Following the description of green fluorescent protein, which was extracted from a jelly fish and has been shown as natural source biomolecules that emit back radiation after excitation

[3] fluorescent proteins have evolved [4, 5]. Several fluorescent biomolecules have been used as a lasing medium in vitro. Cells have recently been studied to be used as the source of laser light, which remained alive even after prolonged lasing action. Such biologic lasing units can be pointed out to be used for intracellular sensing, cytometry and imaging. [6] Light-emitting ability of green florescent protein has been shown as a lasing medium which can absorb blue light and then releasing identical particles of green light [7]. This development can be a promising medical treatment tool to irradiate a tumor with laser emitting cells or to mark them for imaging in the future.

3.2 Stimulated emission of radiation concept

Atoms of an active lasing medium are stimulated via external energy source (excitation mechanism) and its electrons jump-up from their balanced orbital state to a higher orbit by this absorbed incoming energy. Excitation phase help to transit the atoms to higher energy state, which is also unbalanced state (Figure 3b). Those atoms at high energy state return to their "ground state" spontaneously. Then, during transition of those charged electrons to their original orbits (lower energy states), they emit the excess energy as photons (Figure 3c). This emitting process generates identical photons at identical frequency and wavelength. This also provides perfect spatial and temporal harmony (in phase with time and space) of photons. Stimulation of atoms and emission of radiation, which has amplified properties than ordinary light, occurs in the resonator cavity and can be defined as the core of "*Light Amplification by Stimulated Emission of Radiation*-**LASER**" concept.

Wavelength of a laser defines its behavioral characteristics regarding interactions with target materials. Today we have a wide range of laser wavelengths and some are in visible light spectrum. Depending on type of the lasing medium, if the produced laser beam is within the visible light spectrum, its color varies.

3.3 Energy source

Excitation or pumping of active lasing medium is carried out with different methods, so that the atoms could be charged and transported to higher energy levels. As the pumping process start energy loading affects the atoms gradually. Some atoms excited totally, some less-excited and some are in ground sate. As the population of excited atoms is higher than others amplified radiation is emitted. Therefore, lasing medium can't emit radiation until pumping energy level reaches to the excitation threshold. Each lasing medium has its own stimulation method. With this purpose, pumping source should be powerful enough to reach emitting threshold to start working the system shortly after turning on the system, or multiple energy sources can be used simultaneously. This also helps to generate high energy output. Energy pumping systems use different sources in biomedical field.

Electrical

Electric current inside a noble gas medium creates glowing effect, which is also called "Electric glow discharge". It's simply used to pump diode lasers and semiconductor crystal lasers. Similarly free electron lasers and some excimer lasers can be pumped with electron beams.

Optical

Optical pumping is used commonly through the lateral wall of the resonating cavity, which is also termed as "side-pumping". Earliest energy sources are flash lamps (xenon and krypton). Arch lamps (xenon, krypton, argon, neon, and helium) and flash lamps are still two of the most common optical pumping systems in use. Similarly some diode lasers also can be source of optical pumping and excite solid state or liquid dye lasing mediums alone or in combination of another pumping system. Correspondingly, sun beam has also been used as a pumping source of a laser [8]. Microwaves or radiofrequency can be an alternative source of pumping procedure.

Chemical

Chemical reactions can yield energy that can be used for energy pumping. Very high output power (up to megawatt) levels and continuous wave mode can be generated with chemical pumping technique. Chemical oxygen iodine laser (COIL)- 1.315 µm, all gas-phase iodine laser (AGIL)- 1.315 µm, and the hydrogen fluoride laser- 2.7-2.9 µm and deuterium fluoride lasers- 3.8 µm are the examples of this system. Chemical or optical excitation can be used due to high control efficacy on energy pumping. Such high power lasers are mainly used in industry and military.

Gas dynamic excitation systems are used alone or in combination.

Energy pumping mechanism of this system is a thermodynamic process and depends on heat emission of a gas during sudden pressure decrease (adiabatic cooling). Preheated ad pressurized gasses (mostly CO_2) is allowed to expand suddenly into a lower pressured medium with a supersonic velocity. This leads adiabatic cooling, which reduces the temperature and transfer the released energy to the active lasing material.

3.4 Optical resonating cavity

Of the main components of laser system is resonator cavity that surrounds active lasing medium and energy source for pumping. During excitation procedure, especially with high power laser systems, also excessive amount of heat is generated that may rise up to levels that may damage the system. Even low power laser systems may necessitate a cooling system around the lasing medium and energy source.

Photons generated in the cavity travel inside the cavity and reflected by two parallel mirrors one on either pole of lasing cavity, so that they can be reflected in between until a parallel beam profile is yielded (Figure 1). Of one is 100% reflective while the other is 90-95% reflective due to collimation window that let the beam output leaving the cavity.

Since 1960's, geometry of the resonating cavities has been developing, due to the difficulties to produce entirely parallel mirrors. Also convex and concave mirror types have been introduced with the development in optical science and correspondingly cavity designs have evolved to obtain higher beam profile. However, even these advancements on resonating cavity designs or mirror geometries there are limitations for diode lasers due to their small size and production difficulties. [9]

3.5 Cooling system

Present laser production process causes an important side-effect-excessive heating of the optical resonating cavity. Pumping source itself is a primary cause of thermal increase. Additionally, excitement of the active lasing medium can also lead increased resonation of lasing molecules, which means rise of temperature. Temperature of the resonating cavity can reach critical limit during operation, which may easily harm the system and cause severe damages. Therefore, heat control of the laser system mostly inevitable to avoid accidents. Laser cooling systems have been developed depending of the system requirements, while some low-power diode (semiconductor) systems may not necessitate cooling due to their low potential for heat production. More particularly; high repetition rate lasers requires a cooling system. This system generally comprises a closed coolant circulatory loop, which can be viewed in two pieces. First part of the loop is around the laser cavity and an annular coolant heat exchanger forming another portion of the loop. Coolant circulation means including an impeller extending into the loop adjacent an end of the annular heat exchanger are disposed concentrically within the core of the heat exchanger. Coolant absorbing heat generated by laser cavity transfers it to heat exchanger part to discharge externally. Secondary cooling layer can be necessary for high power systems that generate excessive heat.

Gases or liquids can be used as coolant. Water is presented as the more efficient and reliable cooling medium due to its high dissipation rate. However; liquid coolants have become less favorable due to their auxiliary requirements like filters, neutralizers, deionizers and expansion chambers for problem-free circulation in the closed loop environment. Moreover, corrosion and optical degradation associated with the use liquid coolants result in a considerable decrease in reliability and greatly increases requirements for maintenance. Furthermore axial flow impellers which are suitable for use in pumping compressed gases are inherently much more reliable than liquid pumps. Additionally, flashtube replacement in the presence of a liquid coolant is a wet dripping process. Of these issues another one is that excessive water pressure creates vibrations in the laser system.

The heat transport capacity of compressed nitrogen is comparable to that of liquid laser coolants but the effects of temperature extremes such as freezing boiling and expansion associated with liquid coolants are avoided. The compressed gas coolants have proved to be very efficient and therefore have increased greatly the life of the flashtubes.

3.6 Delivery system

Component of a laser system that is responsible to deliver the laser beam to, and/or the receiving the light from the source and transmit to the target is "delivery system". This can be viewed into two parts as: main light carrier and a piece responsible to focus or defocus beam on the target.

a.   Main carrier

During manual operation clinicians require to aim the laser beam to a target area or point for desired shooting. Therefore, besides main working laser system a visible guide laser (mostly low power He-Ne laser) is required to help the user for proper aiming. Main carrier should allow entry and transmission of two different (guide laser and main operational laser) wavelengths simultaneously without any interference. The delivery systems should provide minimum energy output difference between source and tip for maximum power gain.

Most common types of the carrier systems are optical fibers and hollow waveguides. Fiber optic systems are consisted of a bare fiber core surrounded by fused cladding layer basically. Light travels through the lumen at less than a critical angle to be totally reflected whenever it hits the core-clad interface. Internal reflection allows the entire beam energy to be disseminated along the length of the fiber. Fiber systems are highly flexible and transmit radiation over long distances with constant beam diameter and minimal energy loss. This gives the possibility to transfer laser beam into an area, which is difficult to reach without a big access opening. Especially this is very helpful for surgical applications where removal of lesions located in an inner cavity or deep tissue is required. Such interventions (minimally invasive surgery) can be done via endoscopic systems from a small entry hole or without opening a large tissue cut (incision). Choice of fiber diameter can vary from 200µm, 300µm to 600µm for systems used in medical treatments. It should be considered that widening the laser beam to a wider lumen would lead divergence and spreading the output onto a larger spot size. This would be helpful to reduce the power on the target surface. On contrary; using fibers at thinner diameter keeps energy density at high level on the target or allows application of the laser beam into a narrow spaces, like a tooth canal. Though fiber optic systems are capable to transmit laser energy with minimal loss, recent efforts on composition of fiber optic system have been given to improve these features.

Clinician, with fiber systems, is capable to use the beam in contact mode without any applicator device. Operator can touch the fiber tip on target tissue and directly give the energy to the tissue, so the operation can be fulfilled with minimal energy loss and highest efficacy on the target. This application method circumvents spreading of energy to wider tissue surface that permits the surgeon to make an incision with minimal tissue damage.

Hollow waveguide is the alternative light transportation system. Following the beam output is expanded with an upcollimator, laser beam enters in to a hollow tube with reflective inner layer. Straight beam is directed to the target via articulating joints with reflective mirrors. Reflecting mirrors cause loss of power until it reaches to the target. During the travel of laser light in the hollow waveguide it expands and though lenses focusing can be a problem to keep the power loss at minimum level. This delivery system necessitates use of applicators for focusing. CO_2 and Er lasers commonly use this system.

b.   Applicator

Laser beam transmitted via carrier system reaches an applicator (a hand-piece or microscope) that adjust (mostly focus) beam profile and output according to the requirements of application. Applicator can focus the beam at micro scale, so that operations can be performed at the level of a single cell. Under the description of minimally invasive surgery; lasers have in principle the capability to cut a cell or restructuring of target tissue [10] Microbeam allows highly precise noncontact placement of foreign materials into targeted cells without compromising cell viability, which has been extremely difficult task for conventional techniques. Laser assisted delivery of exogenous material into a specific area of three-dimensional tissue, will help to understand the functioning of cellular structure or understand physiologic turn-overs and pathologic changes of a healthy cell [11]. This technique can be beneficial to mark some cells with traceable materials (dye, radioisotope, electromagnetic materials etc.) to follow their signals inside the body to explore their behavior. Exogenous material implantation can be a novel application in treatment of infectious diseases and cancer as well. Drug implanted immune system cells, which are responsible from body defense, can be used in the short future to destroy or mark the diseased area. Moreover; this method can be a promising for genetic studies.

Issues result due to collateral thermal damage to the surrounding living structures and protein denaturation has been overcome with ultrafast (picosecond) shots, so that a clear analysis possibility and problem-free wound healing could be yielded. [12, 13]. Use of such novel applications is highly essential in surgical procedures where sensitivity is required (like ophtamlology or central nervous system surgeries).

It's obvious that working at micro-scale would necessitate microscope use for proper visualization and application features. Therefore, laser systems to be used for micro-scale should be adapted to a microscope system. Some lens and mirror sequences can be used to provide a beam spot-size at micro scale. Micromanipulator devices can also be adapted to work with focused laser beam for precise operational performance. Robotic control feature opens a new era in laser applications with higher safety and more accuracy for the laser procedures that requires repeatability [14]. Automation of laser shooting with micromanipulator system also shortens the operation time and precise execution of surgical plans defined by the surgeon [15].

Major disadvantage with focusing is that the focus distance between the applicator tip and the tissue is constant. To work on a large object with a constant spot-size is almost impossible. A serial production unit it's easy to keep the optics fixed and move the object, however in the surgery room it's not wise. Therefore, surgeon should move the applicator connected to articulated tube system, which is not flexible and difficult to manipulate, to make a clean cut on the tissue. However; under this condition beam divergence and alignment can change and nonhomogenous irradiation profile is created. On the contrary, applicator provides a divergent beam profile to spread the beam on the target so that wider spot-size.

4. Features of laser

4.1 Monocromatic

Day (or ordinary bulb) light is a composition of different colors, which means it contains more than one wavelengths (for us: colors), in other term it's polychromatic (Figure 4). One of the differences from day light is that laser consists of only one wavelength-monochromatic. Regarding the monochromacity; all energy is used to produce one wavelength at high power in a laser system, while same amount of energy is diverted and generates multiple wavelengths (polychromaticity) with ordinary light system. That makes laser energy more profitable to produce one wavelength higher effect on the target and more power with same energy. Also light emitting diodes (LED) are monochromatic.

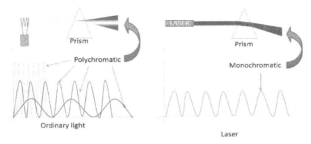

Fig. 4. Comparison of ordinary light and laser regarding mono-, or polychromocity.

4.2 Collimation

Ordinary light produced by bulb or fluorescent tube and light emitting diode (LED) etc. are radiated from one point uniformly in all directions. Some lens and mirror systems have been used to produce parallel beam with limited success and can't overcome divergence issue, while laser systems can produce almost parallel beam profile-**collimated**. (Figure 5) As the divergence ratio of beam rises, power density of beam spreads on a larger target area proportionally (inverse square law).

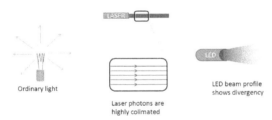

Fig. 5. Illustration shows emitted beam propertes of ordinary light, laser and LED light

Power density (irradiance) is also inversely proportional with square of distance between beam source and target. (Figure 6) Therefore, minimal divergence and use of all energy in one direction in a collimated beam lumen make laser able to transfer majority of output power to the target even at longer distances.

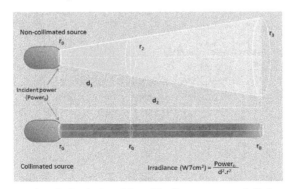

Fig. 6. Irradiace (power density) relationship with divergence and distance.

If needed for special conditions (cutting or drilling materials or tissues), laser beam can be focused with lens systems so hat smaller spot-size can be yielded. Focusing the beam provides higher power density in the projection area on target and makes it possible to work at micro scale. In contrast; divergence can be created intentionally with defocusing lens systems, if needed to irradiate a wider area at single shot. However, it should be kept in mind that irradiating at defocused mode yields lower power density on target. This can be a practical way of reducing power density on the target, especially used in low level laser therapy. Or, defocusing the laser beam can be required to vaporize or process a large area in a short time. Higher power settings are should be used for this purpose.

4.3 Coherent

Photons of generated radiation energy have a releasing sequence and it also determines sequence of collision with the target. When laser is considered, generated photons resonating are synchronized. In other words; laser photons act together in one direction and hit the target simultaneously-coherent. On the other hand; non-laser sources (even LED beam) generate energy pulses sequentially and hit the target asynchronously (Figure 7a,b). That also explains the power of laser beam of which photons hits the target together at the same moment that brings additive effect. This feature of laser loads energy more easily at shorter exposure time and can be more destructive as well.

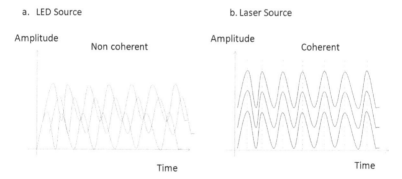

Fig. 7. Comparison of (a.) LED and (b.) laser sources regarding coherency.

5. Modes of operation

Generated laser beam can transfer the energy to the target. Given energy is directly related with the output amplitude and time. Laser systems emit the photons as long as the stimulation energy is given and this creates a continuous beam output, which is also named as **continuous wave (CW)** mode. CW output transfers energy to the target without interruption (Figure 8a) that creates high thermal effect at collision site. It's mostly used to destroy the aimed target by loading excessive energy and rise the temperature until it's burned out. On the other hand beam output can be delivered as interruptedly (**pulsed** mode) (Figure 8b).

Interruption (energy off) periods and pulse (energy on) durations can be modulated up to operator's purpose. Pulsed beam creates beating effect on the target that particularly benefited

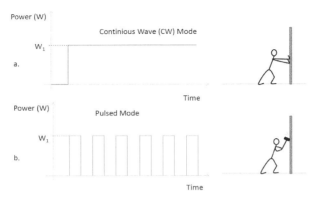

Fig. 8. Working modes of laser. Line graphic show (a.) continuos wave and (b.) pulsed beam profiles.

to destroy more solid materials, like ceramics, metals, bone or enamel of tooth, with less or no thermal effect. Depending on the length of on/off periods (pulse duration/intervals between pulses) transferred energy varies, as well as amplitude (Figure 9).

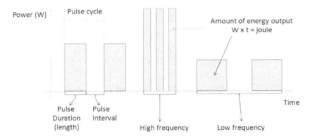

Fig. 9. Transferred energy amount is related with pulse duration and intervals between two pulses. Given energy is related with magnitude of power (W) and pulse length in one pulse cycle, and repetition rate (Hz) in unit time.

Though same amount of energy output is yielded; different pulse durations or repetitions rates affect the target in different ways. Longer pulses act like CW and tend to create more thermal damage on the target and heat it easily; while short pulses create mechanical beat effect with less or minor thermal damage like hammering. With this purpose shorter pulse rates (ultrafast lasers) were introduced at microsecond (10^{-6} second), nanosecond (10^{-9} second), picosecond (10^{-12} second) and femtosecond (10^{-15} second) ranges. Depending on the type of laser system optically methods like mode locking or pumping power modulation (gain switching) or Q switching techniques are used to generate pulse mode operations.

Q (Quality) switch

Q switch is a device placed inside the resonating cavity to prevent light from transmitting outside the resonating cavity basically. It comprises various attenuators to causes very low or rather high losses, respectively, for a laser beam sent through it. This helps to produce short intense pulses with high peak-power. Mostly they are used where the pulse duration is typically in the ultrafast range.

High thermal effect is also used to destroy a tumor tissue or stop bleeding in the operation theater, killing microorganisms or welding of metals in industry but also in tissue closure instead of stiches. Welding technique will be detailed more under "Welding" topic. Additionally, it can be beneficial to heat deeper tissues, of course at lower power settings, intentionally for therapeutic purposes. This stimulates tissue regeneration and blood circulation on the target area, in cases when increased tissue metabolism is aimed to resolve inflammatory conditions [16]. Thus, pulsed energy at high frequencies and at proper amplitudes can breakdown molecular links of a substance in crystal form and may create cracks or even break them apart with micro-explosions. This helps the clinicians to remove hard tissues like bone, enamel or dentin of a tooth for treatment of different conditions.

6. Light-tissue interaction

Interaction phenomenon occurs between at least two entities. Interaction of two entities is a happening depends on natures of two components of interaction (laser and target). Basically, two kinds of interaction can be identified regarding the way of affected side: photon influence on tissue or tissue influence on photons. Regarding these interaction paths uses of electromagnetic energy on organisms can be roughly classified as:

1. Evaluation and research step: In other terms "tissue influence on photons"
2. Impaction step: "Photon influence on tissue" term describes; modification of tissue by laser photons' energy. In other terms; photon bombing modifies physical (photothermal effect) or chemical (photochemical effect) properties of target material. Type and degree of interaction is affected by several factors. power density,
   a. Surgical aim: cutting, evaporation, coagulating, welding and dissecting, at high power settings.
   b. Therapeutic aim: tissue regeneration via inducing ATP synthesis inside the cells, cell replication, blood circulation and, reduction of pain, tissue swelling (edema), etc. at low level power settings.
   c. Production or modification of biomaterials: machining, processing, welding, coating and oxidizing.

Despite plenty efforts are given for the impaction step tissue-photon interactions, evaluation and research step is used to test results of clinical applications or material production steps. Use of biomedical optics and software simulations both for tissue impact and evaluation processes take place within this chapter to guide research scientists, professionals in biotechnology and medical health care.

Those features of laser beam which makes it a new tool in field of medicine. Diagnostic value of laser is also mentioned in this chapter as well as its therapeutic use. The professionals working in biomedical or health sciences will be able to learn about clinical uses of laser systems with clinical cases treated as well as gaining vision for the research and development branch of our science.

In order to understand the way how radiation energy is used in biomedical field; beam-target material (tissue) interaction knowledge should be digested. Understanding the working modes, power density and wavelength variation help to realize what kind of effects laser energy has on the target materials. Regardless the type (wavelength) of laser, when the beam is directed to a tissue, it follows 4 action paths as it touches the surface reflection, transmission, scattering, absorption (Figure 10).

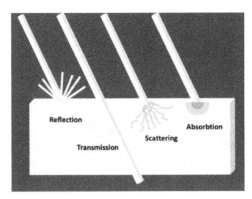

Fig. 10. Tissue interaction of laser beam.

First two types of interactions can be considered as *"tissue influence on photon"* is used for diagnostic sensing, imaging, and spectroscopy of tissues and biomaterials. This step of interaction aims to collect signals released from tissue of echo of sent photons. Those collected signals are processed and analyzed to evaluate the condition on tissue site or examined material or process.

6.1 Reflection

Some part of electromagnetic energy is reflected back by the tissue surface, which is a good example for tissue effect on laser photons. This kind of optical performance is commonly used for diagnostic or imaging studies to keep patient records. This reflected beam is read by a group of sensors and delay for reflection time is used to calculate the distance between the beam source and target point to scan surface morphology (Figure 11).

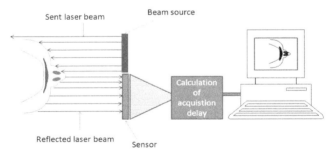

Fig. 11. Laser scanning process. Laser beam is sent to the surface to be scanned and distance between source and arget is calculated with reflection time. Computed raw data helps to reconstruct target surface topography 3 dimentionally.

3D scanning is a core application depending on used many fields of biomedical technologies. Replication of an object depends on scanning part and lasers can give an excellent alternative with high resolution in a short scanning time. Three dimensional laser scanning of macro structures of an organism, like part of (face, hand etc.) or full body [17] is helpful to lineout a topographical profile to measure or diagnose morphological abnormalities. This technique helps the clinician to compare soft tissue topography and landmarks before [18] and after a surgical intervention or to show the patient an estimated treatment outcome before the surgical procedure. Facial templates based on 3D laser face scans can be used to treat some face deformities by guidance of growing skeleton during childhood [19]. Laser surface scanning is also a useful tool to keep records of patients undergoing interventions at face and head.

Three dimensional digital face records obtained from laser scanning can be archived, before and after images can be compared so that success of intervention that affects the face appearance can be visualized. [20, 21] Also face images taken before the intervention can be processed to predict and show the patient end result beforehand [22]. Or image galleries can be a good source to be worked on retrospectively. Similarly dental casts of the patients can be scanned for similar reasons. Also scanned dental models can be superposed to computed tomographic or magnetic resonance imaging data to integrate facial stereophotogrammetry acquisition [23], which helps clinicians to diagnose or treatment planning. Laser scans at micro-scale can be valuable for production of dental restorations. Impressions taken from dental arch can be scanned entirely or partially with laser or ordinary light at micro-scale to produce die used preparation of dental restoration precisely [24].

Currently optoacoustic tomography systems using 2 lasers operating in the near-IR spectral range (755 and 1064 nm) allow 3D imaging of individual organs and blood vessels through the entire body in animal models [25]. This novel method, in the future, would open a new era for imaging for human body.

Besides quantitative static evaluation of the body structure; laser scan systems can aid assessment of dynamic functions of externally, like recording of chest movements during breathing. However, such 4D (time dependent 3D chest position) requires a long acquisition time [26].

Laser has another point of interest in dentistry to differentiate carious tooth. This methodology depends on difference between fluorescence degree of laser beam from health and carious tooth structures [27].

6.2 Transmission

If target tissue is poor of light absorbing ingredient (chromofor), which is responsible to soak the photon energy and stop the beam at superficial layers of the tissue, photons are transmitted to deeper layers until all their energy is lost. Therefore, clinician should be well instrumented concerning which wavelength is absorbed by which chromofor and if the target tissue contains related chromofor or not. Based on this knowledge clinician can yield superficial or deep tissue effect intentionally or accidentally.

Depending on absorption rate of laser energy by the chromofor while travelling inside the tissue, some part of energy, gradually, can pass through all tissue layers entirely and may exit from the other side of tissue. In such cases leaving part of the beam can be used to measure some tissue ingredients via their transmission capability. With this aim the exiting part of beam can be read by a sensor to evaluate absorption rate. Therefore transmission of laser beam is a valuable feature used in diagnostic measuring. Pulse oxymeter devices read transmitted part of beam from the finger-tip and calculate percentage of oxygen concentration in the circulating blood. This is a useful technique to monitor vital signs of a patient closely without damaging any tissue. Similarly, working mechanism of laser doppler flow-meter depends on transmitted laser beam trough a human tooth. This device is used to test if there is a blood circulation inside the tooth pulp. According to this diagnostic value clinician decides the treatment choice.

6.3 Scattering

Photons are spread out their directions during their transmission to the tissue depths, when they hit any reflective structures located in the tissue. This leads change of beam direction during the travel inside the tissue. This may cause defection of effect on desired tissue depth.

6.4 Absorption

Photons transfer their energies to the target substance (chromofor). As a result of this action the target structure gains and transform energy, which may change its physical, chemical structure or increase the temperature. Energy absorption by chromofor occurs in 3 phases. First one is the excitation of atoms. Likewise pumping phase of lasing medium, electrons of chromofor atoms excited. As interaction at atomic level happens with gamma and X-rays, this level is out of focus for this laser chapter. Second level of tissue-radiation interaction is at molecular level, which stays within our biomedical interaction interest. Therapeutic laser applications in medical field are based on this interaction level. Such reactions excite electron bonds of biomolecules, which may break longer molecule bonds or can make changes in carbon chain [16]. On the other hand, biomolecules' oscillation can also be amplified to higher modes, which result as temperature increase macroscopically (photothermal reactions) or rotational stages can be seen in the molecules that result photochemical reactions (photodynamic therapy). Especially when therapeutic applications of laser are concerned those mentioned levels of interactions are our main effect mechanisms inside the tissue. Use of laser in clinic for therapeutic aim; targets to increase blood circulation at the lased area by heating affect and also promotes new material synthesis by the tissue cells. Regarding the therapeutically applications lasers are set at low level power-settings within certain time limits no to exceed to high photothermal reactions. Low-level-laser therapies (LLLT) are predominantly used to cure diseases by promoting tissue regeneration via photochemical activity and inducing blood circulation basically. [16]. However, as higher doses are applied or LLLT is continued over predefined time limits photothermal efficacy of laser beam becomes prominent. Continuing photon energy transfer or applying high doses associate an increase on oscillation (heat) of target molecule. At the beginning, highly oscillating molecules transfer their heat to cooler neighboring areas radially by means of conduction. This is closely depended on the tissue's thermal conductivity and diffusivity. This generated heat (photothermal effect) can be neutralized by this transfer mechanism and it's reversible. If lasing continues or higher doses are applied, tissue conduction can't suffice to transfer all the heat loaded and local temperature reaches to critical limit and cause coagulation of biomolecules that leads death (necrosis) of living cells [28]. Results of this time dependent critical temperature increase are irreversible (Figure 12).

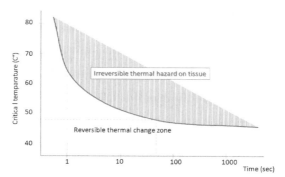

Fig. 12. Time dependent critical temperature limits lead coagulation of biomolecules and tissue necrosis. Data according to Niemz MH and Steiner R.

Further energy loading by prolonged time or with higher power doses; increase local heat until boiling level of water component. Seconds after this stage, water evaporates and the soft tissue loses its volume related with its water component ratio. Especially soft tissues, like mucosa, present a cavitation like crater on the surface of lased area. More lasing burns biomolecules, likewise proteins etc., overlaying the tissue, which causes char formation on the lased surface (Figure 13).

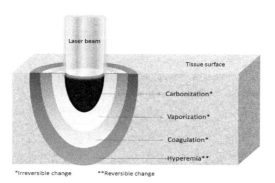

Fig. 13. Laser tissue interaction

Even though, laser tissue interaction mainly depends on the type of laser (wavelength) and also ingredients of tissue interacting with the laser beam; water is most commonly mentioned absorbent of laser in the for wavelengths longer than 1200 nm and shorter than 200 nm. On the other hand, within the visible and near ultra-violet spectrum, water can be considered as transparent and transmitted through the tissue depths. Regarding the tissue composition laser beam may cause deep hazardous effect under such conditions. Therefore, for example, clinician who's going to work with a laser at wavelength between 200 to 1200 nm should know that the beam would easily penetrate deeply a tissue with high water component, like mucosa. Light absorbing organic molecules can be classified into two groups. Of the first contains amino acids and nucleus acid bases and other group the so-called chromophors. Also chromophors can be defined, mostly, as molecular structures which absorb light in the visible spectrum. Line graphic in Figure 14 summarizes atomic absorption co-efficiency over wavelengths of commonly used lasers for biomedical purposes. Based on this graphic clinicians determine which wavelength they should work with according to their purpose.

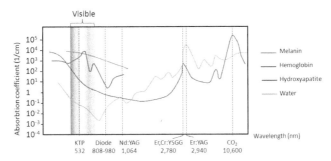

Fig. 14. Absorption spectrum of laser wavelengths.

Regarding this absorption co-efficiency graphic, erbium (Er:YAG and Er:Cr:YSGG) family and CO_2 lasers had a great affinity to water and hydroxyl-apatite crystal and their photons leave their energy on HA and water. This can be an example to choose a proper wavelength to work on tissues with high HA concentration (like bone and tooth) or with high water component like mucosa. Therefore, health care professionals dealing with tooth or bone in their operational practice or researchers etc. should prefer wavelengths between 2780 to 10600nm to obtain highest efficacy. On the other hand, when a high bleeding risk is expected during a soft tissue surgery, a wavelength that is highly absorbed by hemoglobin (blood) should be chosen to be able to stop bleeding.

7. Principals to be acknowledged for dosing

An operator, who has the obligation to work with lasers, should be equipped with basic knowledge of dosing and power-settings. It's necessary to universalize and application and its repeatability. Keeping records of those measures is a must at least to correlate the results with power-settings.

Watt (W) is used as power unit of laser equipment. Given energy (joule-J) is calculated with multiplication of used power (W) and application time (seconds) for continuous wave modes.

Energy (Joule) = power (Watt) x time (seconds)

However, when pulsed systems are regarded; total given energy given in a single pulse (pulse energy - J), repetition rate in a second (Hertz- Hz) and time (s) must be considered as in the following formula:

Energy (J) = pulse energy (J) x repetition rate (Hz) x time (s)

Energy density is termed to calculate the amount of energy delivered over a surface during laser therapy. This term defines the distribution of energy over a surface, so that a routine clinical application of the treatment method can be prescribed for all users as shown below:

Energy density (Fluence) (J/cm2)= power (W) x time (s) / area (cm2)

Clinicians should be founded of all those knowledge in order to prescribe a universal power-setting for clinical application to treat different diseases. Similarly, keeping records related with a treatment operation is important regarding compare healing results and power dosing. So that; in case of a postoperative complication clinician can be able to look back to the records of the operational procedure and may decide to reduce the power settings for the next patient to avoid such unintentional malpractices.

8. Clinical applications

Today, parallel to development of technology in health sciences, radiation has found a wide scope of use for diagnosis and imaging of many diseases and their treatment. Integration of computer and machine engineering to radiation tools spawned a new generation of radiation emission tools with higher capabilities on precise power control. Herewith, selective effectiveness of irradiation on parts of living organism is aimed in current studies. Irradiation is also X-ray, gamma radiation, infra-red, ultra-violet waves, microwaves and ultrasonic energy are also used with this purpose. Radiation energy with its wide use in biosciences has a special importance, however and in spite of such apparently wide acceptance sources for the professionals in bioengineering field is quite limited. The context of second part related with "Clinical applications" aimed to pull out and emphasize the concepts related laser energy used in medical sciences and literature review to purify knowledge for undergraduate, postgraduate biomedical engineers, health care professionals.

Laser systems have been in use in the medical sciences for diagnostic and therapeutic purposes since the 1970s. Clinical applications of laser beam can be explained with "photon influence on tissue" concept basically. Laser energy is used to modify tissue structure both in photochemical and photothermal actions. Depending on energy/power used during medical applications, lasers can be classified into 2 types: high power applications and low-level-laser therapy (LLLT).

8.1 Surgery

8.1.1 Soft tissue cutting (incision)

High power lasers can cut like a knife or ablate (remove large volumes) soft tissues with their photothermal effects. With this respect almost all surgical branches benefit the advantages of laser tool. Most prominent practical applications of lasers are in surgical field due to their heat production capability. Thermal damage capability of beam at desired focus diameter enables high power lasers a preferable surgical tool to cut the tissue precisely at smaller spot-sizes [29]. Especially, when incision on mucosa (rich of water component) is considered erbium family lasers would be more advantageous over other wavelengths due to high specificity to water molecule and low collateral thermal effect [30].

Surgical procedures in ophthalmology necessitate high precision for incision or disruption of transparent tissue at a certain depth. It's been presented that femtosecond lasers with low thermal collateral effect and 1.5microm spot-size can achieve subsurface scleral photodisruption can be achieved in vitro for a variety of intrascleral incisions [31]. Moreover, when precision of incision geometry and depth on a curved eye surface is considered computer guided laser beam with minimal spot size can provide prestigious advancement on surgical practice [32].

Laser cutting of delicate soft tissues has also another focus on laser-assisted microdissection, which will be discussed under research section in this chapter.

8.1.2 Tissue ablation (vaporization)

Surgical procedures targeting removal of a pathologic tissues extending on large tissue surfaces of occupying large tissue volumes, especially with bleeding tendency, can safely be removed with laser ablation quickly [33, 34, 35, 36]. Depending on wavelength selection laser can help the operator to ablate tissue superficially without damaging underlying biomaterial, likewise dental implant exposure lying beneath the oral mucosa (Figure 15).

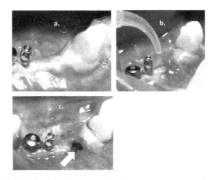

Fig. 15. Tissue ablation with Nd:YAG laser. a. Oral mucosa covering dental implant, b. Mucosa ablation, c. Exposed dental implant (arrow)

Tissue ablation with laser beam is an advantageous technique especially for removal of lesions operated endoscopically and can also provide less bleeding [37]. Owing that the beam focusing ability lasers can compromise application scope like making incision at a small size spot or tissue ablation at larger spot sizes (Figure 16).

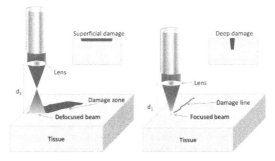

Fig. 16. Focusing and defocusing. Laser beam creates deeper impact at a narrow diameter at surface when focused. If defocused, damage depth is more superficial but damage area on the surface is wider. To defocus the beam $d_1 < d_2$.

Additionally, collateral thermal damage potential of laser ablation technique also widens safety of the operational procedure by killing cells neighboring vaporization center. So that any remnants of malignant tumor cells, if present within the thermal necrosis zone, are killed and any recurrence risk is minimized [38]. Antitumoral action of laser energy can also be achieved photochemichal reactions. This concept will be mentioned within photodynamic therapy section at following context. Like tumor resection in all surgical branches, laser energy has been attributed within the extent of ophthalmology for phacoemulsification of eye lens to be replaced with an artificial one [39].

Among other surgical applications selective and precise tissue ablation is required in ophthalmology. Besides lesion or excessive tissue removals [40]; automated tissue ablations at micron-scale are also very common to modify refractive index of transparent part of eye to correct some visual deficiencies like myopia, hyperopia, and astigmatism [41].

Similarly, dermatology and plastic and reconstructive surgery clinicians take the advantage of laser ablation for removal of several lesions appearing on skin [42]. With cosmetic purposes scar tissues, tattoos and skin discolorations are also removed with lasers. When pigmented lesions are the target to be destroyed with photoablation, laser systems with wavelengths lower than 1064nm are preferred due to their absorption spectrum by pigmented materials. In the present application protocols lasers present successful results and therefore are inevitable for the cosmetic procedures like hair removal and acne treatments. Photothermal and photochemical efficiencies take action simultaneously with these clinical purposes and result least thermal collateral damage on surrounding tissues. Thermal injuries of tissues, especially the skin, would easily cause scarring which leads contraction following tissue healing. In an esthetic point of view tissue contraction is an outcome that can be interpreted with 2 controversial effects. It can be regarded as a negative outcome that may lead uncontrolled scarring with unpleasant esthetic appearance for the patient at first sight. However; on contrary, if laser irradiation procedure can be fulfilled precisely on the skin, the tissue contraction can be an advantageous application to tighten up loosened aging skin [43].

8.1.3 Blood coagulation (hemostasis)

Besides cutting laser with wavelength can easily coagulate the blood and be able to stop blooding simultaneously [44]. That gives a clear vision for the operator especially in body regions like oral cavity (Figure17) with high bleeding tendency.

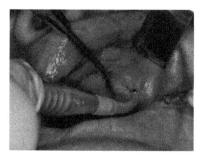

Fig. 17. Use of Nd:YAG laser to stop bleeding with its phothermal efficiency during intraoral soft tissue surgery.

This shortens the operational time for the surgeon, lead less edema and relapse risk [36, 45] and provides a clear vision during the operation. Laser can also be a surgical tool for coagulation solely in association with scalpel surgery. This method is identified as "Laser assisted surgery" concept that can be adapted to almost all surgery disciplines using laser. When vascular pathologies of organ systems are considered, laser system with high coagulation effect can easily help the surgeon to remove vascular pathologies with life-threatening bleeding risk more safely, due to its blood-stopping effect [46-51]. This feature of laser beam makes it a superior tool to scalpel when removal of lesions with high bleeding tendency [52] or surgical approaches of patients receiving anticoagulant therapy [53] or bleeding tendencies [54] are considered.

Lasers carried with fiber optic systems are compatible with endoscopic surgical applications where the surgeon can operate the field with via fiber optic visualization systems. Deep organ systems, like gastrointestinal tract, respiratory system (nose, sinuses, throat, lungs etc.), urinary tract, abdomen, central and peripheral nerve systems, joints and canals of glands can be operated with small entry holes on the skin. Fiber systems can carry laser beam to the target through narrow entry holes without damaging surrounding healthy tissues. Similarly intraocular bleeding, which may cause blindness very easily can be stopped with laser coagulation without touching any tissue [55].

8.1.4 Bone surgery

Though laser has been a focus of many experiments on living organisms since the beginning [56], its potential for bone cutting has been pointed out during 1975 [57]. Since beginning of 1980's, following development of CO_2 [58] and erbium family lasers [59], researchers have widened operational scope of laser systems for bone removal. Even though; erbium lasers looks like similar bone removal capacity as with mechanical instruments, bone healing following removal procedure is still controversial due to collateral heat production with laser beam [60, 61].

8.2 Photodynamic therapy

Photodynamic therapy (PDT) is a technique to sensitize target cells or microorganisms to light source and destroy them more efficiently with photochemical reaction based on photon energy transfer and conversion. Extrinsic dye material (photosensitizer) and intrinsic molecule (oxygen) act as mediators to facilitate the transformation of molecular oxygen from triplet to singlet units via photon energy. Singlet oxygen is highly toxic to the living organisms and destroys the pathologic (tumor) cells, or microorganisms with laser at lower power settings. Although the first article appeared in 1956 [62], it was discovered more than 100 years ago by observing the killing of microorganisms when non-toxic dye (photosensitizing agent) and visible light were combined in vitro. Afterwards, it has become popular technique, due to its low systemic toxicity rate and precise application manner as a treatment for cancer, ophthalmologic disorders and in dermatology. Recently, interest on antimicrobial PDT has evolved and become a focus, due to antibiotic resistance and issues with distribution of antimicrobial drug. It has been indicated as a therapy for a large variety of local infections. Advantages of PDT include equal killing effectiveness regardless of antibiotic resistance, and a lack of induction of PDT resistance. It should be considered that antimicrobial effect would diminish when the light is turned-off. However, on the other hand less than perfect selectivity to microorganisms is a remarkable advantage of antimicrobial PDT. Several dye materials are used to sensitize the microorganisms via light activation. Even use of hemoglobin (blood) as a dye material with antimicrobial PDT over some oral microflora pathogens, like bacteria and fungus, in vivo has been shown to be effective [63]. PDT has been successfully used to kill pathogens and even to save life in several animal models of localized infections such as surface wounds, burns, oral sites, abscesses and the middle ear. A large number of clinical studies of PDT for viral papillomatosis lesions and for acne refer to its antimicrobial effect, but how important this antimicrobial efficacy is unclear for overall therapeutic outcome. Antimicrobial-PDT for periodontitis is a rapidly growing clinical application and other dental applications are under investigation. PDT is being clinically studied for other dermatological infections such as leishmaniasis and mycobacteria. Owing that the rise of antibiotic resistance; antimicrobial PDT applications looks like to have promising future. Advancement on bio-nanotechnology (NT) indicate innovative application spectrum in bioscience. Especially recent researches present similarities in clinical design and mechanistics. NT offers the ability to miniaturize the light source and optimization, enhancing targeting and intensity of the photodynamic reaction as well as a far greater insight into dosimetry and mechanisms of action [64, 65].

8.3 Tissue welding

In surgical branches wound closure has a specific significance, which is directly affect results of wound healing after the intervention. Moreover; some specific fields may require water or air tightness besides simple closure. Conventionally, tissue closure is carried out with simple sutures (stiches), however simple suturing cannot fulfill 100% secure air- or water tightness. Therefore photocoagulation of laser energy had been pointed out as an alternative and superior method for cases where wound sealing is required. Depending on the laser light sources used photothermal effect has another application in surgical practice like adhesion of two soft tissue parts. This is conventional technique to close surgical wound sites without sutures (stiches) based on coagulation of an organic soldering material, which adheres two parts of tissue and hold them together until the tissue heals [66] even in cornea closure [67]. Tissue welding method can bring an innovative ease for the procedures where suturing difficulties are faced, especially in endoscopic interventions [68]. This is a promising technique for blood vessel [69] and nerve bundle [70] anastomosis. Water-tightness of tissue closure has a specific importance for the cardiovascular surgery, but also for neurosurgical procedures. Therefore; laser tissue welding has been studied and results showed that its able to repair cerebro-spinal fluid leaks [71]. Most studies present results with enhanced tissue healing with laser tissue welding, most probably due to its collateral therapeutic effect, series besides anastomosis capability. Besides albumin [72, 73], synthetic polymers are also shown as the source of soldering materials for tissue welding. CO_2 [74] and diode-980 and 1064 nm [75] lasers are most commonly studied systems with successful tissue welding results.

8.4 Therapeutic radiation

Use of laser radiation is not limited to only with destructive applications in medical field. It has been demonstrated that laser energy at low doses has a positive effect on tissue regeneration by increased biomolecule synthesis, formation of connective tissue and increased blood circulation via generation of new vessels [76]. By the way it stimulates tissue healing while helping resolution of inflammation and pain [16]. Basically, photon energy is converted to some important biomolecules like protein, collagen, DNA and/or ATP via chemical reactions and can help to modify cell metabolism. That mechanism can amplify cell replication capability that leads regeneration. Almost all kind of tissue cells can be stimulated for reproduction. Damages due to trauma, infection, etc. can cause tissue loss that can be repaired by low-level-laser therapy (LLLT). Following any injury inflammatory chemical mediators are released by the organism, which cause pain and edema (swelling) [77, 78].

8.5 Dentistry

Besides use of laser energy for soft tissue removal or hemostasis (blood stopping) for oral surgical purposes, it found a wide range of clinical and laboratory application spectrum in field of dental medicine. When the term of "Dentistry" is uttered, first memory reminded is painful drilling with noisy and vibrating rotary tools. However, by the evolution for laser technology hard tissue removal has become more comfortable for the patients. Regarding wavelength absorption graphic in figure 14 erbium lasers have highest absorption rate by water and flowingly by HA. That makes erbium lasers most suitable spectrum for ablation of hard tissues like enamel and dentin without thermal damage (Figure 18).

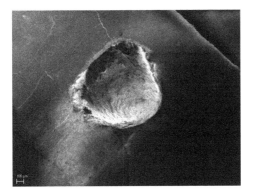

Fig. 18. Scanning electron microscopy shows smooth crater formation on tooth enamel with Nd:YAG laser ablation. (Courtesy of A. Binnaz Hazar Yoruc.)

Non-thermal ablation capability of erbium has made it a novel technology for dental applications like drilling of enamel and dentin [79]. Hard tissue removal with conventional techniques, like rotary instrumentation, looks more faster than ablation with laser; loss of vibration sense and selective ablation make laser more preferable tool than mechanic systems especially in dental procedures when removal of old fillings are considered [80]. Moreover, laser assisted cavity preparation is also advantageous when composite filling materials, which necessitate advanced roughness of the prepared surface, will be utilized [81]. Removal of dentin with laser transmitted via a fiber optic cable helps the dental clinicians for root canal preparation, where antibacterial efficiency is also required [82].

However, low heat production risk of erbium makes it superior regarding thermal necrosis of the living tissues like dental pulp [83] and bone marrow [84].

On the other hand, photothermal efficiency of laser beam is required in some cases like hypersensitivity of dentin [85]. The term of dentin hypersensitivity is used in cases where dentin tissue is exposed external environment due to several reasons and nerve fibers of dental pulp are stimulated by external factors like cold or acidic substances. Laser beam acts on the superficial dentin layer and melt it with its photothermal effect. This melted dentin seals open dentine tubules and blocks communication between exterior and interior environments of a tooth. This obliteration hinders entry of irritants into the dental pulp and overcomes hypersensitivity issue.

When antimicrobial PDT is considered may be dental medicine is the field that most frequently benefits from its advantageous features. Comparing other parts of the body oral cavity has a quite rich microflora regarding variety and quantity of microorganisms. Therefore, oral infections are very common in our mouth. Besides the photothermal efficiency laser energy can kill microorganisms with its antibacterial PDT effect. Laser-induced antibacterial PDT has become a novel method to treat some specific infections were eradication of microorganisms is not possible with chemical or mechanical techniques. Dental implants placed into the jaw bones are produced with a rough surface to enhance implant-bone contact area with increased frictional forces. However, this roughened surface is a proper environment for colonization of the microorganisms, which leads infection of peri-implantar tissues (periinmplantitis). It is highly difficult to eradicate all microorganisms penetrated into the deepest level of the rough surface with chemical and mechanical methods. However, different types of wavelengths can help the clinicians to kill all the bacteria living of the implant surface [86, 87]. Results of the recent studies show that antibacterial PDT is a promising method for the treatment of perimplantitis [88, 89].

Similarly root canal and periodontal pockets of the teeth are the sources of bacteria colonization, which leads loss tooth. With this purpose, antibacterial effect of laser energy looks like a promising tool in management of such endodontic [90] or periodontal infections [91] to provide at least decrease number of pathogenic bacteria in the periodontal pocket or root canals. However, when a periodontal pocket is considered, complete total eradication of microorganism (sterilization) cannot be possible due to irradiated pocket area cannot be isolated from the mouth entirely. Therefore, a disinfection process can be uttered rather than sterilization. Even though reduction of pathogenic bacteria account at the inflammation site co-works with our defense system and help to resolve the local infection.

Several reasons may cause discoloration of teeth that can be attributed as a bad esthetic appearance for some patients. In order to manage this discoloration chemical bleaching of teeth have been a focus of cosmetic dentistry since 1940's [92]. Likewise in all chemical reactions increased heat factor catalyzes the reaction. Researchers realized to benefit converting laser energy to heat and catalyze the bleaching process to shorten the operational time in the office. Therefore, laser systems have become popular in the dental clinics owing that energy transformation to heat induces chemical reaction of hydrogen peroxide during whitening procedure with less regression [93, 94].

9. Laser as a research tool

9.1 Laser spectroscopy

Spectroscopy is an outstanding analytical method, which contributes to present state of atomic and molecular bioscience. This technique uses vibrational, rotational and other low frequency echos in a sample material. It relies on inelastic (Raman) scattering of monochromatic light, mostly a laser beam in the visible, near infrared, or near ultraviolet range. This is a sample for tissue influence on laser photons. The laser interacts with molecular vibrations, photons or other excitation in the system, resulting in an increase or decrease the energy of laser photons. The shift energy provides information about the photon modes in the system. Two of 3 dimensional analysis of any given material even at micro, or nanoscale can be fulfilled with this method [95]. Therefore lasers spectroscopy is an indispensable tool for researchers working all materials science. As a research tool with its high and analytical capability spectroscopy has been used in many fields of bioengineering and medical sciences [96- 98].

9.2 Confocal laser scanning microscopy (CLSM)

CLSM can provide high-resolution optical images at determined tissue depth selectivity. Depending on the laser source and application modes diagnostic data can be yielded via optical information of ultra-high resolution even in deep organ systems via ultra-thin endoscopic probes, which makes confocal laser microscopy inevitable for the physicians who are responsible to diagnose any kind of malignant tumor [99]. So that tissue discrimination and manifold types of fluorescence detection can be achieved. Moreover, optical features would explore 3-D capability of tissues and enhanced recognition of tissue type and pathological status inside the living human body [100]. Early diagnosis of malignant tumors via endoscopy helps the clinicians to save many lives.

It is also called as *optical sectioning*. CLSM images can be yielded layer by layer from deeper levels of tissue to the surface and the data can be reconstructed by a computer to obtain three dimensional topographic appearance of tissue. This can also be a very valuable data for microbiology studies [101].

10. Bioindustry

Likewise in all industrial branches, use of laser technology brought novel advancements occurred in bioindustry as well. Laser systems have taken place in production of complicated or tiny objects with high precision and speed. With this respect laser technology has a special importance both in fabrication and modification of production cascade and repair processes. With material processing regard, increasing demand can be seen in the market for laser systems due to non-contact processing, reduced need for finishing operations, decreased processing cost, high productivity, improved product quality, automation worthiness, greater material utilization and minimum heat affected zone [102]. In biomedical perspective; laser systems adapted to production or repair processes have widen applicability scope and speed with advanced precision and safety. Owing that; production, modification, and assembling can be attributed as the major courses of laser applications in bioindustry.

10.1 Laser – Matter interaction

Regardless the type of procedure, which is being processed; laser-matter interaction, likewise as between laser and tissue should be knowledge. In order not to exceed limits to manufacturing technologies and keep the context as predetermined, rather giving detailed opto-electric physics; briefly the laser-matter interaction can be defined as the same in tissue-laser interaction. In addition to photothermal, photochemical and photomechanic effects laser energy may have photomagnetic effect on materials [25]. Similarly, the interaction is closely related with wavelength – matter reflectivity relation, produced temperature, present surface films, angle of incidence and surface roughness, besides laser beam (power settings, time, shot intervals, focusing, etc.) and matter properties. Laser processing used to be most predominantly preferred in metal processing, however synthetic material processing in bioindustry is also developing with a great acceleration recently, especially in rapid prototyping.

Laser photon impacting the substance surface may or may not change its phase/state. With this regard; phase/state changes can occur in 3 ways:

1. Solids undergo vaporization:
    a. Machining (Drilling, cutting, surface roughening, engraving, scribing, marking etc.)
    b. Coating – deposition (Ceramic coating on metal etc.)
    c. Laser assisted purification
    d. Laser spectroscopy
2. Solids undergo liquid
    a. Adhering (Welding – brazing)
    b. Reclamation
    c. Surface alloying – cladding
    d. Manufacturing (Rapid prototyping – sintering)
3. Liquids undergo solid
    a. Rapid prototyping

On the other hand no phase changes help to modify a solid matter to a desired status as follows:

a. Surface hardening
b. Shocking – shot peening
c. Forming (Bending or straightening)
d. Semiconductor annealing

10.2 Sintering

Laser sintering (LS) has been a focus of bioengineering interest since early 1990's [103]. Basically, a focused laser beam moving on a material (metal, ceramic or polymer) powder surface melts and fuses the powder at a desired shape layer-by-layer. Comparing to conventional casting techniques porosity of the product can be altered as desired with LS. Development of computer technology both in hard- and software contributed to an astounding growth of laser systems. "Three-dimensional" modeling technique adapted to computer guided finely focused laser beam now allows design and production of any object at micro scale. Especially modeling and rapid prototyping concepts have developed within each other based on sintering from a laser perspective. These synergetic developmental changes expand application scope and speed of laser in bioindustry. Rapid fabrication (prototyping) of prosthesis or scaffold [104] with LS before the surgery has 2 major advantages for the surgeon and the patient. First thing first, operator has the ability to work on this study model before the surgery and is capable to prepare any kind of framework or a template, even in complex 3D architecture, to be used as a fixation or stabilization material etc. [105, 106]. On contrary, these time consuming preparation or adaptation steps of the operational procedure should be fulfilled precisely during surgery when tissues are opened and the clock is tick tacking. Secondly, the operator would be familiar with the scene that he/she is going to face during surgery and can prepare and surgical template, which provides a 100% precise bone drilling or cutting capability as planned previously [107].

Those 2 important developments on LS prototyping can be attributed as innovative features that rise success rates of many surgical procedures in maxillofacial surgery, orthopedics [108], neurosurgery [109]. Also bony defects to be repaired with a bone graft replacement can be imagined initially with computed tomography. Advanced softwares enable the operator to visualize and measure the defect area and create a suitable 3D prosthesis to fit defect site perfectly and work as a graft material for replacement. Created 3D model in the digital environment (mostly prepared as a *.stl file) can be transferred to a rapid prototyping laser machine to materialize solid graft material. Related with features of prototyping method computer guided laser beam produce the design from solid materials, powder or liquid feeds. Manufactured bone scaffolds made of osteoconductive materials can serve as a tray for cell cultures with bioengineering applications [110]. LS metals products, like dental crowns, provide superior bonding to ceramic, as the bonding capability is closely related with chemical structure on the metal surface rather than porosity value [111].

10.3 Machining

Laser technology has brought a fresh breath to manufacturing of metal instruments, prosthesis of fixation devices etc. requiring precise cutting, drilling, engraving, labeling or marking. Conventional techniques using mechanical removal of matter result thermal or physical damage of the processed material that may disrupt its biologic performance. In such cases high power lasers, especially with femto- or picosecond shots, can offer an excellent procedure with minimal deformation and chemical damage on the processed material.

10.3.1 Drilling, cutting

Under machining concept material processing drilling and cutting procedures is the two most common ones covered. With this respect highest precision is obtained for drilling, cutting etc. procedures with minimum crust and wastage, or no foreign body remnants [112], which do affect tissue reaction to the product if it's going to be placed into a living body. With this regard ultra-short laser pulses have a wide application scope in production of metals, ceramics and organic materials with enhanced precision with minimal hazard even on highly fragile materials like glass [113].

10.3.2 Surface modification

Final stages of bioindustrial production process may necessitate surface modification of a material at the end for several purposes like labeling, marking, coloring, engraving a specific surface topography (microstructuring), surface hardening and coating the material with another substance for environmental tissue adaptation or at least for marketing. Moreover, laser ablation has been shown as a useful technique to strip (clean-up) a thin layer from the surface of a metal for preparatory processes for further surface modification.

10.4 Microtexturing

In last 3 decades bioindustry has shown a growing interest in microtexturing (structuring) of implant surfaces to improve their physical and biologic characteristics. Amount of contact area between bone tissue and implant surface is highly important for the stability, which can be attributed as the health and survival of implant. Therefore, in such cases surface roughness is increased to gain maximum frictional forces in between. Conventional chemical (acid etching) and mechanical (sand blasting) erosive techniques leaves unintentionally foreign contaminants on the implant surface that may cause rejection of implant by the tissue following an adverse tissue reaction. With this perspective, clean surface roughening even at micro scale at a predetermined surface topography can be available with laser systems (Nd:YAG, CO_2, and excimer) with no foreign substance contamination. Use of ultra-short pulses is indicated as a cost-efficient method manufacturing of bio-friendly medical implant surfaces. [114] Nowadays, researches are concentrated on nano-roughness concept instead of microtexturing [115, 116] due to this nano-roughness can present more attractive physical topography for attachment of different cell types selectively [117]. That laser induced biotechnology would bring the option to modify the implant surface selectively depending on the environment to be faced, like to be an attractive surface for tissue cells, while surface structure can be changed to protect microorganism adhesions.

10.5 Surface coating/cladding - Alloying

Coating a surface has plenty of additive effects for biomaterials like improving biocompatibility or corrosion resistance, surface roughness and chemical bonding capability, therefore is a frequently subjected to many research studies [118]. Laser coating looks like a promising method especially for composite material coatings even at ultra-thin layers on biomaterials that would face complex tissue interactions selectively in bioengineering applications [119]. Under this title; alloying with other metals likewise coating (cladding) with different substances can be uttered, however surface alloying can offer superior corrosion and wear protective coating nature when compared with ceramic coatings.

10.6 Surface heat treatment - Hardening

Photothermal effect of laser helps to treat surface of materials with heat, which helps hardening on metals. This treatment modality used to manage wear reduction; however it has currently found a place in production technologies to change metallurgical and mechanical properties as well. Hardness, fatigue time and strength of metals can be augmented while wear and friction features are reduced with laser heat treatment, if needed selectively.

10.7 Coloring, labeling/marking

Like surface hardening procedures, laser energy can modify metal surfaces' composition and change its color. Many industrial manufacturers practice this technique for classification or marking their products. Color coding is a user friendly technique preferred by manufacturers that help management or classification of hand instruments or drills. Similarly, simple marks showing length printed with lasers on gauges or drills (Figure 19) provides a striking ease for the users to avoid complications during surgery.

Fig. 19. Surface modification with laser on drill for labeling and marking (Courtesy of Bicon Co.)

10.8 Welding

In contrast to conventional welding techniques laser welding makes it possible to assemble two metallic pieces with/without a soldering material remotely and can provide highest heat source intensity on a very small spot size [25]. Therefore, its joining efficiency can be attributed to be higher than other welding systems. Though high costs, this technique can be automatized at atmospheric conditions with high precision with very short on/off timing and cause minimal or no contamination and collateral damage. Materials like magnetic metals of heat sensitive substances can be welded without any damage and shrinkage. Therefore, besides manufacturing of instruments laser welding has been used dental laboratories to joind dental prosthesis parts on standard master models [121]. Excimer, Nd:YAG and carbonmonoxide lasers are more advantageous due to the shorter wavelength natures over the CO_2 laser. Highly focused beam profile minimizes collateral thermal increase in neighboring areas or tissues [122]. Specially designed laser tips (Figure 20) contribute to focus condense the beam in a small spot widens safety limits and opens a new area for laser welding to be performed inside the living organism, like metal welding in mouth [123].

Fig. 20. Specially designed Nd: YAG laser tip for intraoral metal welding. (Courtesy of Prof. Carlo Forniani)

11. Laser safety

Lasers can potentially damage any part of our body unintentionally. Range of those accidents is quite wide, between increased local blood flow at the target area (hyperemia) to death. Therefore; regardless the reason use of laser necessitates some strict safety regulations. Such regulations are issued by local authorities and owner of the laser system has the primary responsibility to operate the system according to those regulations. International standards like ANSI 136.1, 3, 4, 5, 6, 7 and IEC 60825 define classification of lasers according to their hazardous risks and operational powers. Moreover; additional local regulations, if present, should also be obeyed by the personnel dealing with laser operation. First step of these laser safety precautions is training of professionals, who are going to work in an operation environment where a laser works. Regardless their acts in the operation room all personnel should have this basic training. People who are going to operate with laser or laser technicians or laser safety operators responsible from maintenance and management of laser should take advance training [124].

Concerning the damage risk and power of lasers those systems classify lasers Class 1, 1M, 2, 2M, 3R, 3B, 4. Class 3B and 4 are defined as hazardous for eyes respectively. Therefore, safety precautions are at the highest level when other classes are regarded. When eye safety is regarded; wearing goggles with wavelength specific filter is a must within the operation room. Protective precautions like automatic door lock installation at the operation room, isolating untargeted areas, placing warning signs outside the door and "laser safety training" etc are inevitable. Class 3B and 4 class lasers are commonly used for medical purposes for tissue removal or modification, making a cut (incision), eradicating bacteria, killing tumor cells, etc., or biomedical research, or in bioindustry. Clinicians operate laser must have "Laser Safety Training for Healthcare Professionals" certificate. However, health care professionals working in laser operation site have the obligation to get orientation training at least.

12. Acknowledgement

Author would like to acknowledge Prof. Carlo Forniani, Assoc. Prof. A. Binnaz Hazar Yoruc and Bicon Co. for sharing their image data with courtesy.

13. References

[1] Licht SH (1983) Therapeutic Electricity and Ultraviolet Radiation. Baltimore: Williams & Wilkins.
[2] Ford A, Morris SM, Coles HJ (2006) Photonics and lasing in liquid crystals. Materials Today 9(7-8): 36-42.
[3] Shimomura O, Johnson F H, Saiga Y (1962) Extraction, purification and properties of aequorin, a bioluminescent protein from luminous hydromedusan, Aequorea J Cell Comp Physiol (59): 223–239.
[4] Shaner NC, Steinbach PA, Tsien RY. (2005). A guide to choosing fluorescent proteins. Nature Methods. 2: 905–909.
[5] Giepmans BNG, Adams SR, Ellisman MH, Tsien RY (2006) Review—the fluorescent toolbox for assessing protein location and function. Science 312: 217–224.
[6] Gather MC, Yun SH (2011) Single-cell biological lasers. Nature Photonics. 5: 406–410.

[7] Shen Z, Burrows PE, Bulović V, Forrest SR, Thompson ME (27 June 1997) Three-Color, Tunable, Organic Light-Emitting Devices. Science. 276(5321): 2009-2011.

[8] De Young R J, Weaver WR (August 18, 1986) Low-threshold solar-pumped laser using C2F5I. Applied Physics Letters. 49(7): 369–370.

[9] Siegman AE (Nov/Dec 2000) Laser Beams and Resonators: Beyond the 1960s. IEEE J. 6(6): 1389 – 1399.

[10] Rabien A. (2010) Laser Microdissection. Methods in Molecular Biology. 576: 39-47.

[11] Gu L, Mohanty SK (2011 Dec) Targeted microinjection into cells and retina using optoporation. J Biomed Opt. 16(12): 128003.

[12] Amini-Nik S, Kraemer D, Cowan ML, Gunaratne K, Nadesan P, Alman BA, Miller RJ (2010 Sep 28) Ultrafast mid-IR laser scalpel: protein signals of the fundamental limits to minimally invasive surgery. PLoS One. 5(9)pii: e13053.

[13] Amini-Nik S, Kraemer D, Cowan ML, Gunaratne K, Nadesan P, Alman BA, Miller RJ. (2010 Sep 28) Ultrafast mid-IR laser scalpel: protein signals of the fundamental limits to minimally invasive surgery. PLoS One. 5(9). pii: e13053.

[14] Wong YT, Finley CC, Giallo JF 2nd, Buckmire RA (2011 Aug) Novel CO2 laser robotic controller outperforms experienced laser operators in tasks of accuracy and performance repeatability. Laryngoscope.121(8):1738-42.

[15] Mattos LS, Caldwell DG, Dellepiane M, Grant E (2010) Design and control of a robotic system for assistive laser phonomicrosurgery. Conf Proc IEEE Eng Med Biol Soc. 2010: 5411-5.

[16] Baxter GD (April 4, 1994) Therapeutic Lasers: Theory and Practice. Edinburgh: Churchill Livingstone.

[17] Bretschneider T, Koop U, Schreiner V, Wenck H, Jaspers S (2009 Aug) Validation of the body scanner as a measuring tool for a rapid quantification of body shape. Skin Res Technol. 15(3):364-9.

[18] Toma AM, Zhurov A, Playle R, Ong E, Richmond S (2009 Feb) Reproducibility of facial soft tissue landmarks on 3D laser-scanned facial images. Orthod Craniofac Res. 12(1):33-42.

[19] Kau CH, Zhurov A, Richmond S, Cronin A, Savio C, Mallorie C (2006 Feb) Facial templates: a new perspective in three dimensions. Orthod Craniofac Res.9(1):10-7.

[20] Shim WH, Yoon SH, Park JH, Choi YC, Kim ST (2010 Dec) Effect of botulinum toxin type A injection on lower facial contouring evaluated using a three-dimensional laser scan. Dermatol Surg. 36(4): 2161-6.

[21] Shimomatsu K, Nozoe E, Ishihata K, Okawachi T, Nakamura N (2012)Three-dimensional analyses of facial soft tissue configuration of Japanese females with jaw deformity - A trial of polygonal view of facial soft tissue deformity in orthognathic patients. J Craniomaxillofac Surg. 2011 Nov 9. [Epub ahead of print]

[22] Plooij JM, Maal TJ, Haers P, Borstlap WA, Kuijpers-Jagtman AM, Bergé SJ (2011 Apr) Digital three-dimensional image fusion processes for planning and evaluating orthodontics and orthognathic surgery. A systematic review. Int J Oral Maxillofac Surg. 40(4): 341-52.

[23] Rosati R, De Menezes M, Rossetti A, Sforza C, Ferrario VF (2010 Jul) Digital dental cast placement in 3-dimensional, full-face reconstruction: a technical evaluation. Am J Orthod Dentofacial Orthop. 138(1): 84-8.

[24] Chan DC, Chung AK, Haines J, Yau EH, Kuo CC (2011 Sep-Oct) The accuracy of optical scanning: influence of convergence and die preparation. Oper Dent. 36(5): 486-91.

[25] Brecht HP, Su R, Fronheiser M, Ermilov SA, Conjusteau A, Oraevsky AA (2009 Nov-Dec) . Whole-body three-dimensional optoacoustic tomography system for small animals. J Biomed Opt. 14(6): 064007.

[26] Catanuto G, Patete P, Spano A, Pennati A, Baroni G, Nava MB (2009 Nov-Dec) New technologies for the assessment of breast surgical outcomes. Aesthet Surg J. 29(6): 505-8.

[27] Rando-Meirelles MP, de Sousa Mda L (2011 Mar) Using laser fluorescence (DIAGNOdent) in surveys for the detection of noncavitated occlusal dentine caries. Community Dent Health. 28(1): 17-21.

[28] Markolf HN. (2011) Laser-Tissue Interactions: Fundamentals and Applications In: Raulin C, Karsai S, editors. Springer, Kunzi-Rapp K. and Steiner R., Intense Pulsed Light Technology, Laser and IPL Technology in Dermatology and Aesthetic Medicine. Berlin Heidelberg: Springer-Verlag. pp. 37 – 40.

[29] Xia SJ (2009 May)Two-micron (thulium) laser resection of the prostate-tangerine technique: a new method for BPH treatment. Asian J Androl. 11(3): 277-81.

[30] Ryu SW, Lee SH, Yoon HJ (2011) A comparative histological and immunohistochemical study of wound healing following incision with a scalpel, CO(2) laser or Er,Cr:YSGG laser in the Guinea pig oral mucosa. Acta Odontol Scand. 2011 Dec 12. [Epub ahead of print]

[31] Sacks ZS, Kurtz RM, Juhasz T, Mourau GA (2002 Jul) High precision subsurface photodisruption in human sclera. J Biomed Opt. 7(3): 442-50.

[32] Nubile M, Carpineto P, Lanzini M, Calienno R, Agnifili L, Ciancaglini M, Mastropasqua L (2009 Jun) Femtosecond laser arcuate keratotomy for the correction of high astigmatism after keratoplasty. Ophthalmology. 116(6): 1083-92.

[33] Tasar F, Sener BC (2009) Laser use in Oral and Maxillofacial Surgery. In: LASERS in Dentistry: Practical Textbook. Vitale M, Caprioglio C, editors. Bologna: Edizioni Martinada, pp. 243-252.

[34] Santos NR, Aciole GT, Marchionni AM, Soares LG, dos Santos JN, Pinheiro AL (2010 Oct) A feasible procedure in dental practice: the treatment of oral dysplastic hyperkeratotic lesions of the oral cavity with the CO2 laser. Photomed Laser Surg. 28 Suppl 2: S121-6.

[35] Gao H, Ding X, Wei D, Cheng P, Su X, Liu H, Zhang T. (2011 Dec) Endoscopic management of benign tracheobronchial tumors. J Thorac Dis. 3(4): 255-61.

[36] Saccomandi P, Schena E, Di Matteo FM, Pandolfi M, Martino M, Rea R, Silvestri S (2011 Aug) Laser Interstitial Thermotherapy for pancreatic tumor ablation: Theoretical model and experimental validation. Conf Proc IEEE Eng Med Biol Soc. 2011: 5585-8.

[37] Virgin FW, Bleier BS, Woodworth BA (2010 Jun) Evolving materials and techniques for endoscopic sinus surgery. Otolaryngol Clin North Am. 43(3): 653-72, xi.

[38] Ishii J, Fujita K, Munemoto S, Komori T (2004 Feb) Management of oral leukoplakia by laser surgery: relation between recurrence and malignant transformation and clinicopathological features. J Clin Laser Med Surg. 22(1): 27-33.

[39] Chambless WS (1988 Mar) Neodymium:YAG laser phacofracture: an aid to phacoemulsification. J Cataract Refract Surg. 14(2): 180-1.

[40] Lindfield D, Ansari G, Poole T (2012 Feb) Nd:YAG Laser Treatment for Epithelial Ingrowth After Laser Refractive Surgery. Ophthalmic Surg Lasers Imaging. 9: 1-3.

[41] Manche EE, Carr JD, Haw WW, Hersh PS (1998 Jul) Excimer laser refractive surgery. West J Med. 169(1): 30-8.

[42] Goldberg DJ (2008) Laser Dermatology: Pearls and Problems. Massachusetts: Blackwell. pp. 172-4.

[43] Goldman A, Wollina U, de Mundstock EC (2011 May) Evaluation of Tissue Tightening by the Subdermal Nd: YAG Laser-Assisted Liposuction Versus Liposuction Alone. J Cutan Aesthet Surg. 4(2): 122-8.

[44] Yang SS, Hsieh CH, Lee YS, Chang SJ (2012) Diode laser (980 nm) enucleation of the prostate: a promising alternative to transurethral resection of the prostate. Lasers Med Sci. 2012 Jan 27. [Epub ahead of print]

[45] Raewyn C, Paul W (2010 Jun) Management of congenital lingual dermoid cysts. Int J Pediatr Otorhinolaryngol. 74(6): 567-71.

[46] Taşar F, Tümer C, Sener BC, Sençift K (1995 Jul-Sep) Lymphangioma treatment with Nd-YAG laser. Turk J Pediatr. 37(3): 253-6.

[47] Lee JC, Kim JW, Lee YJ, Lee SR, Park CR, Jung JP (2011 Aug) Surgical management of auricular infantile hemangiomas. Korean J Thorac Cardiovasc Surg. 44(4): 311-3.

[48] Daramola OO, Chun RH, Kerschner JE (2012 Jan) A case of large bladder hemangioma successfully treated with endoscopic yttrium aluminium garnet laser irradiation. Arch Otolaryngol Head Neck Surg. 138(1): 72-5.

[49] Takemoto J, Yamazaki Y, Sakai K (2011 Dec) Treatment of Lymphangioma with CO2 Laser in the Mandibular Alveolar Mucosa. Int J Urol. 18(12): 854-6.

[50] Arslan A, Gursoy H, Cologlu S (2011 Nov) Treatment of Lymphangioma with CO2 Laser in the Mandibular Alveolar Mucosa. J Contemp Dent Pract. 12(6): 493-6.

[51] Rameau A, Zur KB (2011 Sep) KTP laser ablation of extensive tracheal hemangiomas. Int J Pediatr Otorhinolaryngol. 75(9): 1200-3.

[52] Hochman M, Adams DM, Reeves TD (2011 May-Jun) Current knowledge and management of vascular anomalies: I. Hemangiomas. Arch Facial Plast Surg. 13(3): 145-51.

[53] Chung DE, Wysock JS, Lee RK, Melamed SR, Kaplan SA, Te AE (2011 Sep) Outcomes and complications after 532 nm laser prostatectomy in anticoagulated patients with benign prostatic hyperplasia. J Urol. 186(3): 977-81.

[54] Green D (2007 Jun) Management of bleeding complications of hematologic malignancies. Semin Thromb Hemost. 33(4): 427-34.

[55] Gil AL, Azevedo MJ, Tomasetto GG, Muniz CH, Lavinsky J (2011 Sep-Oct) Treatment of diffuse diabetic maculopathy with intravitreal triamcinolone and laser photocoagulation: randomized clinical trial with morphological and functional evaluation. Arq Bras Oftalmol. 74(5): 343-7.

[56] Goldman L, Shumrick DA, Rockwell RJ Jr, Meyer R (1968 Mar) The laser in maxillofacial surgery. Preliminary investigative surgery.. Arch Surg. 96(3): 397-400.

[57] Verschueren RC, Koudstaal J, Oldhoff J (1975 Mar)The carbon dioxide laser; some possibilities in surgery. Acta Chir Belg. 74(2): 197-204.

[58] Gertzbein SD, deDemeter D, Cruickshank B, Kapasouri A (1981) The effect of laser osteotomy on bone healing.. Lasers Surg Med. 1(4): 361-73.

[59] Nelson JS, Yow L, Liaw LH, Macleay L, Zavar RB, Orenstein A, Wright WH, Andrews JJ, Berns MW (1988) Ablation of bone and methacrylate by a prototype mid-infrared erbium:YAG laser. Lasers Surg Med. 8(5): 494-500.

[60] Martins GL, Puricelli E, Baraldi CE, Ponzoni D (2011 Apr) Bone healing after bur and Er:YAG laser ostectomies. J Oral Maxillofac Surg. 69(4): 1214-20.

[61] Cloutier M, Girard B, Peel SA, Wilson D, Sándor GK, Clokie CM, Miller D (2010 Dec) Calvarial bone wound healing: a comparison between carbide and diamond drills, Er:YAG and Femtosecond lasers with or without BMP-7. Oral Surg Oral Med Oral Pathol Oral Radiol Endod. 110(6): 720-8.

[62] Schultz KH, Wiskemann A, Wulf K (1956) Clinical and experimental studies on photodynamic efficacy of phenothiazine derivatives with special reference to megaphen. Arch Klin Exp Dermatol. 202(3): 285-98.

[63] Meral G, Tasar F, Kocagöz S, Sener C (2003) Factors affecting the antibacterial effects of Nd:YAG laser in vivo. Lasers Surg Med. 32(3): 197-202.

[64] Allison RR, Mota HC, Bagnato VS, Sibata CH (2008 Mar) Bio-nanotechnology and photodynamic therapy--state of the art review. Photodiagnosis Photodyn Ther. 5(1): 19-28.

[65] Li WT (2009 Oct) Nanotechology-based strategies to enhance the efficacy of photodynamic therapy for cancers. Curr Drug Metab. 10(8): 851-60.

[66] Abergel RP, Lyons RF, White RA, Lask G, Matsuoka LY, Dwyer RM, Uitto J (1986 May) Skin closure by Nd:YAG laser welding 1986 May.

[67] Rasier R, Ozeren M, Artunay O, Bahçecioğlu H, Seçkin I, Kalaycoğlu H, Kurt A, Sennaroğlu A, Gülsoy M (2010 Sep) Corneal tissue welding with infrared laser irradiation after clear corneal incision. Cornea.29(9): 985-90.

[68] Bleier BS, Cohen NA, Chiu AG, O'Malley BW Jr, Doghramji L, Palmer JN (2010 May-Jun) Endonasal laser tissue welding: first human experience. Am J Rhinol Allergy. 24(3): 244-6.

[69] Esposito G, Rossi F, Matteini P, Puca A, Albanese A, Sabatino G, Maira G, Pini R (2011 Apr-Jun) Present status and new perspectives in laser welding of vascular tissues. J Biol Regul Homeost Agents. 25(2): 145-52.

[70] Bloom JD, Bleier BS, Goldstein SA, Carniol PJ, Palmer JN, Cohen NA (2012 Jan) Laser facial nerve welding in a rabbit model. Arch Facial Plast Surg. 14(1): 52-8.

[71] Bleier BS, Palmer JN, Sparano AM, Cohen NA (2007 Nov) Laser-assisted cerebrospinal fluid leak repair: an animal model to test feasibility. Otolaryngol Head Neck Surg. 137(5): 810-4.

[72] Ware MH, Buckley CA (2003) The study of a light-activated albumin protein solder to bond layers of porcine small intestinal submucosa. Biomed Sci Instrum. 39: 1-5.

[73] Pabittei DR, Heger M, Beek JF, van Tuijl S, Simonet M, van der Wal AC, de Mol BA, Balm R (2011 Jan)Optimization of suture-free laser-assisted vessel repair by solder-doped electrospun poly(ε-caprolactone) scaffold. Ann Biomed Eng. 39(1): 223-34.

[74] Youssef TF, Ahmed MR, Kassab AN (2010) Utilization of CO(2) laser for temporal fascia graft welding in myringoplasty: an experimental study on guinea pigs. ORL J Otorhinolaryngol Relat Spec. 72(2): 119-23.

[75] Hu L, Lu Z, Wang B, Cao J, Ma X, Tian Z, Gao Z, Qin L, Wu X, Liu Y, Wang L (2011 Mar) Closure of skin incisions by laser-welding with a combination of two near-

infrared diode lasers: preliminary study for determination of optimal parameters. J Biomed Opt. 16(3): 038001.

[76] Prabhu V, Rao SB, Chandra S, Kumar P, Rao L, Guddattu V, Satyamoorthy K, Mahato KK (2012 Feb) Spectroscopic and histological evaluation of wound healing progression following Low Level Laser Therapy (LLLT). J Biophotonics. 5(2): 168-84. doi: 10.1002/jbio.201100089.

[77] Yan W, Chow R, Armati PJ (2011 Jun) Inhibitory effects of visible 650-nm and infrared 808-nm laser irradiation on somatosensory and compound muscle action potentials in rat sciatic nerve: implications for laser-induced analgesia. J Peripher Nerv Syst. 16(2): 130-5.

[78] Marcos RL, Leal Junior EC, Messias Fde M, de Carvalho MH, Pallotta RC, Frigo L, dos Santos RA, Ramos L, Teixeira S, Bjordal JM, Lopes-Martins RÁ (2011 Nov-Dec) Infrared (810 nm) low-level laser therapy in rat achilles tendinitis: a consistent alternative to drugs. Photochem Photobiol. 87(6): 1447-52.

[79] Rizcalla N, Bader C, Bortolotto T, Krejci I (2012 Feb) Improving the efficiency of an Er:YAG laser on enamel and dentin. Quintessence Int. 43(2): 153-60.

[80] Chan KH, Hirasuna K, Fried D (2011 Sep) Rapid and selective removal of composite from tooth surfaces with a 9.3 μm CO2 laser using spectral feedback. Lasers Surg Med. 43(8): 824-32. doi: 10.1002/lsm.21111.

[81] De Moor RJ, Delme KI (2010 Apr) Laser-assisted cavity preparation and adhesion to erbium-lased tooth structure: part 2. present-day adhesion to erbium-lased tooth structure in permanent teeth. J Adhes Dent. 12(2): 91-102.

[82] Moshonov J, Sion A, Kasirer J, Rotstein I, Stabholz A (1995 Feb) Efficacy of argon laser irradiation in removing intracanal debris. Oral Surg Oral Med Oral Pathol Oral Radiol Endod. 79(2): 221-5.

[83] Kilinc E, Roshkind DM, Antonson SA, Antonson DE, Hardigan PC, Siegel SC, Thomas JW (2009 Aug) Thermal safety of Er:YAG and Er,Cr:YSGG lasers in hard tissue removal. Photomed Laser Surg. 27(4): 565-70.

[84] Stübinger S, Nuss K, Pongratz M, Price J, Sader R, Zeilhofer HF, von Rechenberg B (2010 Oct)Comparison of Er:YAG laser and piezoelectric osteotomy: An animal study in sheep. Lasers Surg Med. 42(8): 743-51.

[85] Sgolastra F, Petrucci A, Gatto R, Monaco A (2011 Mar) Effectiveness of laser in dentinal hypersensitivity treatment: a systematic review. J Endod. 37(3): 297-303.

[86] Stubinger S, Etter C, Miskiewicz M, Homann F, Saldamli B, Wieland M, Sader R (2010 Jan-Feb) Surface alterations of polished and sandblasted and acid-etched titanium implants after Er:YAG, carbon dioxide, and diode laser irradiation. Int J Oral Maxillofac Implants. 25(1): 104-11.

[87] Gonçalves F, Zanetti AL, Zanetti RV, Martelli FS, Avila-Campos MJ, Tomazinho LF, Granjeiro JM (2010 Apr) Effectiveness of 980-mm diode and 1064-nm extra-long-pulse neodymium-doped yttrium aluminum garnet lasers in implant disinfection. Photomed Laser Surg. 28(2): 273-80.

[88] Cavus O (2011) Potasyum titanil fosfat laserin porfiromonas gingivalis ve osteoblastlar üzerine fotodinamik etkinliğinin incelenmesi, Ph.D. Thesis. Marmara University Institute of Health, Istanbul, Turkey.

[89] Shibli JA, Martins MC, Theodoro LH, Lotufo RF, Garcia VG, Marcantonio EJ (2003 Mar) Lethal photosensitization in microbiological treatment of ligature-induced peri-implantitis: a preliminary study in dogs. J Oral Sci. 45(1): 17-23.

[90] Peters OA, Bardsley S, Fong J, Pandher G, Divito E (2011 Jul) Disinfection of root canals with photon-initiated photoacoustic streaming. J Endod. 37(7): 1008-12.

[91] Benhamou V (2009 Apr) Photodisinfection: the future of periodontal therapy. Dent Today. 28(4): 106, 108-9.

[92] Grogan DF (1946 Mar) Agents used in bleaching teeth. Tufts Dent Outlook. 20: 20-3.

[93] Torres CR, Barcellos DC, Batista GR, Borges AB, Cassiano KV, Pucci CR (2011 May) Assessment of the effectiveness of light-emitting diode and diode laser hybrid light sources to intensify dental bleaching treatment. Acta Odontol Scand. 69(3): 176-81.

[94] Al Quran FA, Mansour Y, Al-Hyari S, Al Wahadni A, Mair (2011 Winter) Efficacy and persistence of tooth bleaching using a diode laser with three different treatment regimens. Eur J Esthet Dent. 6(4): 436-45.

[95] Demtröder W (2002) Laser Spectroscopy: basic concepts and instrumentation. Berlin: Springer-Verlag

[96] Baker R, Matousek P, Ronayne KL, Parker AW, Rogers K, Stone N. 2007 Jan) Depth profiling of calcifications in breast tissue using picosecond Kerr-gated Raman spectroscopy. Analyst. 132(1): 48-53.

[97] Samuels AC, DeLucia FC Jr, McNesby KL, Miziolek AW (2003 Oct) Laser-induced breakdown spectroscopy of bacterial spores, molds, pollens, and protein: initial studies of discrimination potential. Appl Opt. 42(30): 6205-9.

[98] Bazalgette Courrèges-Lacoste G, Ahlers B, Pérez FR (2007 Dec 15) Combined Raman spectrometer/laser-induced breakdown spectrometer for the next ESA mission to Mars. Spectrochim Acta A Mol Biomol Spectrosc. 68(4): 1023-8.

[99] Paull PE, Hyatt BJ, Wassef W, Fischer AH (2011 Oct) Confocal laser endomicroscopy: a primer for pathologists. Arch Pathol Lab Med. 135(10): 1343-8.

[100] Stepp H, Sroka R (2010 Oct) Possibilities of lasers within NOTES. Minim Invasive Ther Allied Technol. 19(5): 274-80.

[101] Senges C, Wrbas KT, Altenburger M, Follo M, Spitzmüller B, Wittmer A, Hellwig E, Al-Ahmad A (2011 Sep) Bacterial and Candida albicans adhesion on different root canal filling materials and sealers. J Endod. 37(9): 1247-52.

[102] Majumdar JD, Manna I (2003) Laser processing of materials. Sadhana (Academy Proceedings in Engineering Sciences), 28 (3-4): 495-562.

[103] Bartels KA, Bovik AC, Crawford RC, Diller KR, Aggarwal SJ (1993) Selective laser sintering for the creation of solid models from 3D microscopic images. Biomed Sci Instrum. 29: 243-50.

[104] Niino T, Hamajima D, Montagne K, Oizumi S, Naruke H, Huang H, Sakai Y, Kinoshita H, Fujii T(2011 Sep) Laser sintering fabrication of three-dimensional tissue engineering scaffolds with a flow channel network. Biofabrication. 3(3): 034104.

[105] Torres K, Staśkiewicz G, Śnieżyński M, Drop A, Maciejewski R (2011 Feb) Application of rapid prototyping techniques for modeling of anatomical structures in medical training and education. Folia Morphol (Warsz). 70(1): 1-4.

[106] Suzuki M, Ogawa Y, Hagiwara A, Yamaguchi H, Ono H (2004) Rapidly prototyped temporal bone model for otological education. ORL J Otorhinolaryngol Relat Spec. 66(2): 62-4.

[107] Williams JV, Revington PJ (2010 Feb) Novel use of an aerospace selective laser sintering machine for rapid prototyping of an orbital blowout fracture. Int J Oral Maxillofac Surg. 39(2): 182-4.

[108] Schrank ES, Stanhope SJ (2011) Dimensional accuracy of ankle-foot orthoses constructed by rapid customization and manufacturing framework. J Rehabil Res Dev. 48(1): 31-42.

[109] Stoodley MA, Abbott JR, Simpson DA (1996 Apr) Titanium cranioplasty using 3-D computer modelling of skull defects. J Clin Neurosci. 3(2): 149-55.

[110] Bukharova TB, Antonov EN, Popov VK, Fatkhudinov TKh, Popova AV, Volkov AV, Bochkova SA, Bagratashvili VN, Gol'dshtein DV (2010 Jul) Biocompatibility of tissue engineering constructions from porous polylactide carriers obtained by the method of selective laser sintering and bone marrow-derived multipotent stromal cells. Bull Exp Biol Med. 149(1): 148-53.

[111] Iseri U, Ozkurt Z, Kazazoglu E (2011) Shear bond strengths of veneering porcelain to cast, machined and laser-sintered titanium. Dent Mater J. 30(3): 274-80.

[112] Dausinger F, Lichtner F, Lubatschowski H (2004) Femtosecond Technology for Technical and Medical Applications. Berlin Heidelberg: Springer-Verlag.

[113] Esser D, Rezaei S, Li J, Herman PR, Gottmann J (2011 Dec 5) Time dynamics of burst-train filamentation assisted femtosecond laser machining in glasses. Opt Express. 19(25): 25632-42.

[114] Erdoğan M, Öktem B, Kalaycıoğlu H, Yavaş S, Mukhopadhyay PK, Eken K, Özgören K, Aykaç Y, Tazebay UH, Ilday FÖ (2011) Texturing of titanium (Ti6Al4V) medical implant surfaces with MHz-repetition-rate femtosecond and picosecond Yb-doped fiber lasers. Optics Express 19(11): 10986-10996

[115] Li Y, Gao Y, Shao B, Xiao J, Hu K, Kong L (2012 Jan 16) Effects of hydrofluoric acid and anodised micro and micro/nano surface implants on early osseointegration in rats. Br J Oral Maxillofac Surg. [Epub ahead of print]

[116] Aboushelib M, Salem N, Abotaleb A, Abd El Moniem N (2011 Sep 9) Influence of surface nano-roughness on osseointegration of zirconia implants in rabbit femur heads using selective infiltration etching technique. J Oral Implantol.. [Epub ahead of print]

[117] Nevins M, Kim DM, Jun SH, Guze K, Schupbach P, Nevins ML (2010 Jun) Histologic evidence of a connective tissue attachment to laser microgrooved abutments: a canine study. Int J Periodontics Restorative Dent. 30(3): 245-55.

[118] Pang W, Man HC, Yue TM (15 January 2005)Laser surface coating of Mo–WC metal matrix composite on Ti6Al4V alloy. Materials Science and Engineering. 390(1–2): 144–153.

[119] Prosecká E, Buzgo M, Rampichová M, Kocourek T, Kochová P, Vysloužilová L, Tvrdík D, Jelínek M, Lukáš D, Amler E (2012) Thin-layer hydroxyapatite deposition on a nanofiber surface stimulates mesenchymal stem cell proliferation and their differentiation into osteoblasts. J Biomed Biotechnol.2012: 428503.

[120] Steen MW, Mazumder J (2010) Laser Material Processing. London: Springer.

[121] Apotheker H, Nishimura I, Seerattan C (1984) Laser welded vs. soldered non precious alloy dental bridges: a comparative study. Laser Surg Med 4: 207–213.

[122] Fornaini C, Bertrand C, Rocca JP, Mahler P, Bonanini M, Vescovi P, Merigo E, Nammour S (2010 Jul) Intra-oral laser welding: an in vitro evaluation of thermal increase. Lasers Med Sci. 25(4): 473-7.

[123] Fornaini C, Vescovi P, Merigo E, Rocca JP, Mahler P, Bertrand C, Nammour S (2010 Mar) Intraoral metal laser welding: a case report. Lasers Med Sci. 25(2): 303-7.

[124] Henderson R, Schulmeister K. (November 1, 2003) Laser Safety. New York, London: Taylor & Francis.

4

Biomedical Instrument Application: Medical Waste Treatment Technologies

Muhammed Gulyurt
Fatih University
Turkey

1. Introduction

According to the Ministry of Environment of Turkish Republic *(CYGM),*(2005) medical waste comes out of the premises whose activities result in waste production, such as university hospitals and clinics, general purpose hospitals and clinics, maternity hospitals and clinics, military hospitals and clinics, health centers, medical centers, dispensaries, ambulatory medical centers, morgues and postmortem examination centers, vivisectionist entities, nursing homes and old-age asylums, medical and biomedical laboratories, veterinary hospitals, blood banks and transfusion centers, emergency and first aid centers, dialysis centers, rehabilitation centers, biotechnological laboratories and institutes, medical research centers, other units of health service (doctor offices, mouth and dental health surgeries and so on), veterinary clinics, acupuncture centers, physiotherapy centers, domiciliary treatment and nursing services, centers of beauty, ear-piercing and tattoo, pharmacies, ambulance services and zoological gardens. [1]

In addition, WHO emphasizes that at the domestic level, municipalities, some entities, private companies or organized non-governmental organizations carry out the collection and processing of medical waste through some necessary measures taken by healthcare organizations. Unless the microorganisms contained by the medical waste are collected and made ineffective in a controlled manner, firstly, the medical, then it becomes a threat against public health.[2] Through countries which can not dispose of the hospital waste satisfactorily, some certain diseases are frequently observed in society. Open wound infections, typhoid, gastric, diarrhea, cholera, Hepatitis B and Hepatitis C and HIV threatens mainly doctors' health, nurses and patient careers.[3] It is largely estimated that 21 million people are exposed to HBV, 2 million people to HCV, and 260,000 people to HIV viruses due to the repeated use of unsterilized injections. According to the WHO report, some of the re-use disposable syringes countries are African, Asian, and Central and Eastern European countries. [2]

However, considering the risks of contamination of extremely dangerous communicable viruses, such as Hepatitis B and HIV, through humans, some regulations and laws are put into effect in order to take some precautions necessary for the control and disposal in the premises which produce waste, to process the medical waste separate from others by preventing any harm against public health and environment. Those regulations describe the overall procedure in a detailed way, from the identification of medical waste to its collection and disposal.

Moreover, European Union waste management strategy Council Directive 75/442/EEC, [4] dated 15 July 1975 was repealed on 5 April 2006; and it has been determined to the European Parliament and Council Directive 2006/12/EEC, [5]. In the abovementioned directive, there are some definitions of waste, and classifications of the waste categories included in the directive, and outlines of disposal and recycling activities. The appendix of the directive comprises some provisions which constitute the foundations and main principles of the waste management regarding the waste types outside the coverage, general principles with respect to the waste management, formation of the authorities, the waste management plans, the liabilities of the enterprises which are involved in the recycling and disposal activities, their obligation to hold records and their periodical auditing, the waste disposal expenses covered within the scope of the polluter pays principle and producer responsibilities.

Furthermore, the Commission Resolution 2000/532/EC (2000)[6] underlined that the principles of compiling a common list for waste and hazardous waste to be used in community, in compliance both with Waste Framework Directive and the Council Directive 91/689/EEC, [7] on the hazardous waste. The Waste Framework Directive and the directive of waste list are both adopted by the member countries of European Union, and incorporated into "the directives to be primarily adjusted" [7] sections in the countries in the European Union accession process, such as Turkey. While the regulations and laws vary at the level of states in the USA, it is determined in some international congresses and seminars that they are one step ahead of the European countries.

2. The historical background in the USA

The classification of medical waste was one of the most important questions in terms of the U.S. regulations during 1970s. Its officially first recognition as a distinct waste type was provided by a federal agency in 1978. EPA's approach to the question was one step further in the sense that it principally described infectious waste, i.e. medical waste, under the category of hazardous waste in compliance with the Resource Conservation Recovery Act (RCRA). Thus, although the EPA's proposition for hazardous waste regulations includes infectious wastes as proclaimed in RCRA in 1980, the agency refused that medical wastes did not constitute any serious threat to public health and environment.

Looking at the implementations in European countries and the USA, it can be observed that there are numbers of systems of the medical waste disposal. Besides within the scope of the sterilization methods as the primary Medical Waste Treatment Technologies, there are oxygenic, ozonic and oxygen-free methods of disposing which are preferred in large integrated plants. Although burning is exactly a disposal method, it is not preferred by some countries since a considerable amount of hazardous soot and ash comes out of the burning operation. Expenditures seem to be important in the choice of Medical Waste Treatment Technology methods. Yet, the resolutions taken by the governments and the laws to which they are subjected are directing factors in their choices. The purpose of this chapter is to analyse the analyze the Medical Waste Treatment Technologies.

3. What are the hazardous medical wastes

Medical waste is defined by the World Health Organization as "a broad range of materials such as used needles and syringes to soiled dressings, body parts, diagnostic samples, blood, chemicals, pharmaceuticals, medical devices and radioactive materials which are produced health care clinics, centers, hospitals and laboratories."[8]

Through European Union Legislation 75/442/EEC [4] dated 15 July 1975 defines the meaning of "waste", while the Council Directive describes "hazardous waste" in the European Union Legislation 91/689/EEC [7] on 12 December 1991. Within the European Union Legislations, the descriptions of the types of hazardous wastes, the precautions to be made, the waste management, the transmission and disposal of hazardous wastes are available. The countries in the European Union accession process, on the other hand, exercise those regulations in the period grated in accordance with the course of the process.

3.1 International biotechnology logo

The biohazard symbol which is associated with the term "biological hazards" simply means that it is extremely important to take precautions due to a potential exposition of the materials threatening public heath. The Dow Chemical Company first designed the symbol for their waste storages in 1966.[9]

Fig. 1. Biohazard warning symbol. The sign color stipulated in the standard form is fluorescent orange-red. [9]

3.2 What is the medical waste categories?

The European Union Council Directive 67/548/EEC [10] was promulgated on 27 June 1967, with respect to the adjustment of laws, regulations and administrative provisions regarding the classification, packaging and labeling of hazardous waste. Based on this directive, the Ministry of Environment and Forestry of Turkish Republic published "the Regulations for the Medical Waste Control 25883" on July 22, 2005, [1] and classified the wastes collected from hospitals and health care organizations as follows.

Classification of the wastes stemming from health care organizations						
Household wastes (20 03* and 15 01*)		Medical wastes (18 01* and 18 02*)			Hazardous wastes	Radioactive wastes
A: General Wastes 20 03*	B: Packing Wastes 15 01 01*, 15 01 02*, 15 01 04*, 15 01 05*, 15 01 06*, 15 01 07*	C: Infectious Wastes 18 01 03* and 18 02 02*	D: Pathological Wastes 18 01 02*	E: Sharp Wastes 18 01 01* and 18 02 01*	F: Hazardous Wastes 18 01 06*, 18 01 08*, 18 01 10*, 18 02 05*, 18 02 07*	G: Radioactive Wastes
Parts with the presence of healthy people, sections in which those who are not patient are treated, first aid areas, administrative units, cleaning services, kitchens, wastes coming from storehouses and workshops: any wastes arising from medical centers excluding those mentioned in the groups B, C,D,E, F and G,	Reusable and recyclable wastes coming from all administrative units, kitchens, storehouses, workshops, etc. : -paper -cardboard -paperboard -plastic -glass -metal, etc.	Wastes which requires special treatment for their transmission and disposal in order to prevent the expanding of infectious agents: Microbiological laboratory wastes Cultures and stocks Infectious body liquids Serologic wastes Other contaminated laboratory wastes (lamina, lamella, pipette, petri, etc.) Blood, blood products and any materials contaminated with those Used surgical clothes (fabric, gowns, gloves, etc.) Dialysis wastes (sewage and equipments) Quarantine wastes Air filters contaminated with bacteria and viruses Experimental animal carcasses, organ parts, their blood, and any materials touched by those	Anatomical tissue wastes, parts of organs and bodies along with body liquids coming out during a medical treatment such as surgery, autopsy, etc.:: body parts, organic parts, placenta, and separated organs, etc. stemming from surgery rooms, morgues, autopsies, forensic medicine (humane pathological wastes) Guinae pig carcasses used in biological experiments	Wastes which may cause pinprick, sharp, scar and any injuries: injectors, other cutters with syringe, bistoury, lamina-lamella, glass pasteur pipette, other broken glasses, etc.	Wastes to be subject to special treatment due to their physical or chemical qualities, or because of any legal causes: -Dangerous chemicals -Cytotoxic and cytostatic drugs -Amalgame wastes -Genotoxic and cytotoxic wastes -Pharmaceu-tical wastes -Wastes comprising heavy metals -Pressured containers	They are collected and removed in compliance with the regulations of Atomic Energy Institution of Turkey

* European Union European Waste Catalogue Code Numbers

"The Regulation for Medical Waste Control" (22 June 2005/25883). Table 1 Part I [1]

Wastes originated from health care organizations	European Union European Waste Catalogue Code Numbers
Household Wastes	20 03 01
Packaging Wastes	15 01 01, 15 01 02,15 01 04,15 01 05, 15 01 06, 15 01 07
Infectious Wastes	18 01 03, 18 02 02
Pathogenic Wastes	18 02 01
Sharp Wastes	18 01 01, 18 02 01
Hazardous Wastes	18 01 06, 18 01 08, 18 01 10, 18 02 05, 18 02 07

Table 1. Part II, (2001/118/EC) [12]

3.3 Damages caused by medical waste

According to the U.S. Ocean Dumping Act which was promulgated in 1972 for the prohibition of intentional hazardous waste disposal into the sea, the penalties can be prison sentence for 5 years and/or pecuniary punishment up to $125,000 for civil violations as well as up to $250,000 for criminal offences. Obviously, this falls short of preventing or discouraging the offenders from polluting the sea with medical wastes, leaving used materials and other hazardous wastes which pose a serious threat against public and environmental health. [13]

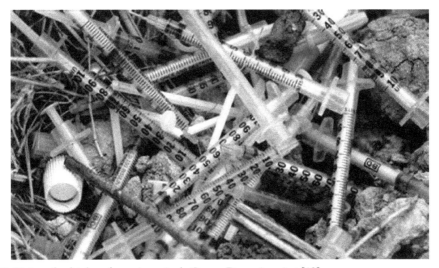

Pic. 1. Picture had taken from America's Ocean Dumping Act [13]

Unless the medical wastes are made ineffective in an appropriate manner during their collection, anyone present at that time is open to plausible risks Medical wastes are left in empty places without any processing in some countries. Public health is threatened by viruses unless they are destroyed by animals and ground waters.

Doctors and nurses who work at the premises which produce medical waste as a result of their activities constitute the groups at risk as they are in touch with the patients' blood and body liquids.

3.4 Medical waste risk group

Due to unsatisfactory medical waste managements, certain groups are open to risk or prone to threat:

- Doctors, nurses, and other healthcare personnel
- Those who attend in the premises which produce infectious waste, such as patients, and hospital attendants,
- Those who work at the waste processing establishments
- Scavengers that feed on waste yards or non-isolated collection points

Children who are potentially in touch with hazardous waste, used materials, etc. [2]

4. Accidents caused by sharp materials

Medical wastes may comprise pathogenic viruses. As a result of inappropriate collection, those viruses may cause infection through open wounds on the body, or, any other ways on the skin which are not deep but with a high level contamination, such as mucous membrane and respiratory passage. Medical wastes should be collected in the bags with the biohazard logo separate from household wastes so that viruses do not threat human health as a result of simple mistakes.

The Table 2 shows the infection types, how they infect and sample causative organisms. The infection types caused by blood and other body liquids (saliva-mucus, urine, tears, etc.) are clearly indicated. There are necessary precautions to stop the spreading of infection in hospitals, to protect people from the infections stemming from the patients during the treatment process.

Type of infection	Examples of causative organisms	Transmission vehicles
Gastroenteric infections	Enterobacteria, e.g. *Salmonella*, *Shigella* spp.; *Vibrio cholerae*; helminths	Faeces and/or vomit
Respiratory infections	*Mycobacterium tuberculosis*; measles virus; *Streptococcus pneumoniae*	Inhaled secretions; saliva
Ocular infection	Herpesvirus	Eye secretions
Genital infections	*Neisseria gonorrhoeae*; herpesvirus	Genital secretions
Skin infections	*Streptococcus* spp.	Pus
Anthrax	*Bacillus anthracis*	Skin secretions
Meningitis	*Neisseria meningitidis*	Cerebrospinal fluid
Acquired immunodeficiency syndrome (AIDS)	Human immunodeficiency virus (HIV)	Blood, sexual secretions
Haemorrhagic fevers	Junin, Lassa, Ebola, and Marburg viruses	All bloody products and secretions
Septicaemia	*Staphylococcus* spp.	Blood
Bacteraemia	Coagulase-negative *Staphylococcus* spp.; *Staphylococcus aureus*; *Enterobacter*, *Enterococcus*, *Klebsiella*, and *Streptococcus* spp.	Blood
Candidaemia	*Candida albicans*	Blood
Viral hepatitis A	Hepatitis A virus	Faeces
Viral hepatitis B and C	Hepatitis B and C viruses	Blood and body fluids

Table 2. Examples of infections caused by exposure to health-care wastes, causative organisms, and transmission vehicles, Report of a WHO Study Waste from health-care activities [14]

There are chiefly four viral diseases in case of any contact with patients' blood and contaminated body liquids. Those are HIV, Hepatitis B (HIB), Hepatitis C (HCV) and Hepatitis D (HDV) viruses. In the chapter published by WHO, Table 3 shows the number of people who are injured by the sharps among those who work at the waste management and in the U.S. hospitals and the number of HBW through injuries. The HBV infection can easily catch through those who are once exposed to blood and other infectious blood liquids and tissues.

Professional category	Annual number of people injured by sharps	Annual number of HBV infections caused by injury
Nurses		
in hospital	17 700–22 200	56–96
outside hospital	28 000–48 000	26–45
Hospital laboratory workers	800–7 500	2–15
Hospital housekeepers	11 700–45 300	23–91
Hospital technicians	12 200	24
Physicians and dentists in hospital	100–400	<1
Physicians outside hospital	500–1 700	1–3
Dentists outside hospital	100–300	<1
Dental assistants outside hospital	2 600–3 900	5–8
Emergency medical personnel (outside hospital)	12 000	24
Waste workers (outside hospital)	500–7 300	1–15

Table 3. Viral hepatitis B infections caused by occupational injuries from sharps (USA) [8]

The risk of contamination increases as a result of inappropriate storage; unless the living creatures who share the same habitat, such as cats, dogs, birds, flies, mice and insects are kept far away from those areas, they can conduct microbiological pathogens and infect humans. An inappropriate medical waste storage system results in the expanding of pathogenic viruses and the transformation of medical waste into an unhealthy environmental problem. In some countries, medical wastes are collected together with household wastes; this gives rise to unhealthy scenes with which street animals can easily get in contact as it can be seen in the following pictures.

Pic. 2. (Picture had taken in 1995) Medical Waste Management in the Hazardous Waste Group, the Report of the Deputy Chair of the Environment Commission of the Grand National Assembly of Turkey [15]

The instructions published in different countries suggest that especially sharp wastes should be conserved in plastic boxes instead of bags. As every country has different application criteria, the entities which produce the waste are notified about necessary precautions by the relevant units.

As it could be seen below, sharp materials are put in plastic bags, which in turn, pose a big threat for the environment and human life because they can contact directly, with patient body liquids and easily spread viruses and pathogens as a result of this contact. As it is obvious in the picture below, syringe, sharp apparatuses are put into inappropriate bags, usually plastic bags is preferred and those bags could be easily torn during their collection or removal and they poese a huge threat for the environment.

Pic. 3. Medical Waste Management in the Hazardous Waste Group, the Report of the Deputy Chair of the Environment Commission of the Grand National Assembly of Turkey [15]

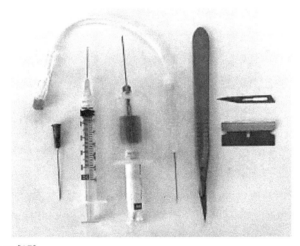

Pic. 4. Sharp objects [15]

Sharp objects should be conserved in the hard surface plastic boxes after they are used. As it can be seen in the Picture 5-6, instead of the bags with the possibility to be torn or punctured during transmission, it is considered more preferable to use plastic boxes which can not be torn or punctured.

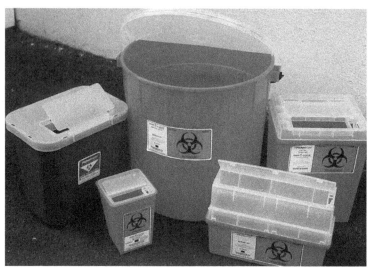

Pic. 5. Princeton University (2008), "Laboratory Waste Streams" picture available at [Ref, 16]

Pic. 6. Hazardous Waste Management Plan (HWMP), (1990), California Country of El Dorado, available at [Ref 17]

4.1 Methods of treating hazardous medical waste

There are tons of medical wastes collected from hospitals. There are also different methods applied by certain institutions, municipalities or firms which work in compliance with the related authorities to treat this hazardous medical waste.. Money spent for the application of such methods is the main criteria in choosing these methods. In addition, in some countries, medical waste is disposed in an environment-friendly manner.

- Containment processes
i. Landfilling in municipal disposal sites

Pic. 7. UNEP (United Nations Environment Programme), (2007), "Waste management is problematic throughout Sudan" [18]

Pic. 8. UNEP (United Nations Environment Programme), (2007), "Waste management is problematic throughout Sudan" [18]

Unsanitary disposal which gives rise to environmental pollution seriously threatens both living creatures and human health. Considering the damages which come out of inappropriate medical waste storage, it is prohibited to bury infectious wastes in regular storage areas, in compliance with European Union Directive 99/31/EC. [19] In the non-member states, this application causes serious harms against environment and human health. Viruses contaminated in the buried medical wastes threaten human health through the conductor insects and the ground waters being mixed up with drinkable water. It is also inevitable that those who share the same living area with animals and plants infect humans through food chain. (Pic 9) The medical wastes buried without any processing or dumped in an uncontrolled way give damages to the environment as potential microbe nidus.

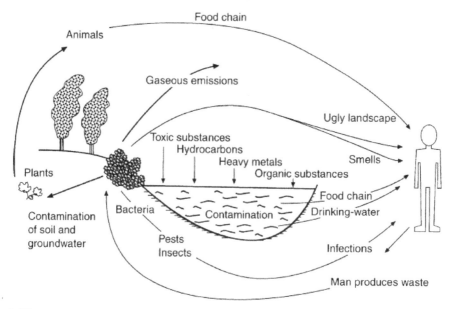

Pic. 9. [20]

Medical wastes should be filled on an impervious land without being pressed. It is required to cover the waste over after daily storage, and this should be covered by shielding with 30 cm earth. In doing so, sanitary waste storage becomes applicable between the impervious land and the upper clay lump ground. Any living creature will not bee able to reach at the medical waste covered by earth, and it will be prevented for surface waters to penetrate. This method which is prohibited in European countries is still being preferred by some countries since it is the most economic method in comparison with other disposal methods. Sanitary medical waste storage method is presented in the Picture -12-. For sanitary storage, the ground should be first made impervious. It is necessary to prevent the leakage stemming from the buried medical waste. (Drainage coefficient is available in the directives of countries. Any drainage system should not be built smaller than k= 1×10^{-4}.)

Pic. 10. Medical Waste Management in the Hazardous Waste Group, the Report of the Deputy Chair of the Environment Commission of the Grand National Assembly of Turkey [15]

4.2 Incineration technologies (medical waste incineration)

While disposal of medical waste through incineration is one of the most reliable technologies, it costs extremely high to establish a plant which will harm the environment at a minimum level. The first investment granted for incineration systems and the cost of plants are much higher than other technologies.

Pic. 11. *"Open burning of medical waste may not destroy pathogens and produces considerable toxic emissions"* (Batterman S. 2008)[21]

If medical wastes are incinerated as in the Picture 13, they give much more serious damages to the environment. It is especially necessary for medical wastes not to comprise materials with heavy metals, radioactive wastes, broken thermometers or other materials with mercury and cadmium.

The flue gas emission values should be definitely kept under control because the gas which appears during the incineration operation causes serious damages against environment and penetrates into the atmosphere that we breath.

The carbon monoxide which appears at hight levels in the flue gas emission values indicates the incineration operation is cariied out at low temperatures. The unburned hydrocarbons, dioxin and furans, volatile heavy metals and their compunds, and other gases such as TOK, HCI, HF, SO2, NOX comes out of this operation. Those gases should be exhausted to the atmosphere through the flue after they are falled under the determined standards. Those flue values are declared by the relevant authorities of countries; and those institutions are auditting the plants which dispose medical wastes with a granted license. Emission guidelines for "hospital/medical/infectious waste" incinerators below Table 4.

Pollutant	Small incinerator ($\leq$91 kg/hour)	Medium incinerator (>91–227 kg/hour)	Large incinerator (>227 kg/hour)
A. Emission limits for new incinerators (construction after June 1996)			
Particulate matter	115 mg/m^3	69 mg/m^3	
Carbon monoxide (CO)	40 ppmv	40 ppmv	
Dioxins/furans	125 ng/m^3 total CCD/CDF or 2.3 ng/m^3 TEQ	125 ng/m^3 total CCD/CDF or 2.3 ng/m^3 TEQ	125 ng/m^3 total CCD/CDF or 2.3 ng/m^3 TEQ
Hydrogen chloride (HCl)	100 ppmv or 93% reduction	100 ppmv or 93% reduction	100 ppmv or 93% reduction
Sulfur dioxide (SO$_2$)	55 ppmv	55 ppmv	55 ppmv
Nitrogen oxides	250 ppmv	250 ppmv	250 ppmv
Lead	1.2 mg/m^3 or 70% reduction	1.2 mg/m^3 or 70% reduction	1.2 mg/m^3 or 70% reduction
Cadmium	0.16 mg/m^3 or 65% reduction	0.16 mg/m^3 or 65% reduction	0.16 mg/m^3 or 65% reduction
Mercury	0.55 mg/m^3 or 85% reduction	0.55 mg/m^3 or 85% reduction	0.55 mg/m^3 or 85% reduction

Sources: Safe Management of Wastes from Health-care Activities' which was published by WHO in 1999 [8]

Table 4.

Whille the post-incineration energy can be used as heating energy in winter times, it can also be used in the power generation in summers. While the energy recovery plants which convert those wastes into energy serve the disposal of medical wastes which threaten human life, it is for good that those wastes are being recovered as power or heat necessary for human life. In developed countries, the double stage systems are used, the most common of which is Controlled-Air Incinerators.

As the most common medical waste incineration (MWI) system, Controlled-air incineration is predominantly set up at health care centers and other medical institutions. It is also known as starved-air incineration, two-stage incineration, or modular combustion. A typical schematic diagram of a controlled air unit is represented in the Picture 12

There are mainly two stages of the waste combustion in this mechanism. In the first stage, waste is fed into the primary, or lower, combustion chamber, which is operated with less than the stoichiometric amount of air required for combustion. Combustion air which is also called as primary or under fire air enters the primary chamber from beneath the incinerator hearth (below the burning bed of waste). In the primary (starved-air) chamber, the low air-to-fuel ratio dries and facilitates

Volatilization of the waste and most of the residual carbon in the ash burns. Under these circumstances, combustion gas temperatures are relatively low (760 to 980°C [1,400 to 1,800°F]).

While the organic substances in the medical waste are being burned, the pathogenic organisms are also disposed. After the incineration the medical waste decrease by 95% in volume, and by 80% in mass. The duration of the medical wastes held in the incineration unit varies in relation to the types of systems.

The post-incineration (volatile) gases accompanied with excess air are sent to the second stage so that they can be burned completely. Secondary chamber temperatures are higher than primary chamber temperatures—at least from 980 to 1,095°C (1,800 to 2,000°F). These values are changeable, sometimes needed to increase, according to the heating value and moisture content of the waste through auxiliary burners located at the entrance to the secondary (upper) chamber to reach at the desired temperatures. This operation enables the demolition of organic toxic substances such as dioxin and furan. There might be a considerable between the minimum and the maximum waste feed capacities for controlled air incinerators. For example, it ranges from about 0.6 to 50 kg/min (75 to 6,500 lb/hr) at an assumed fuel heating value of 19,700 kJ/kg [8,500 Btu/lb]. On the other hand, lower heating value wastes require a higher degree throughput capacity because feed capacities are limited by primary. The unit size and options purchased are the determining factors on whether waste feed and ash removal will be carried out manually or automatically. [22]

The post-incineration (volatile) gases accompanied with excess air are sent to the second stage so that they can be burned completely.

Secondary chamber temperatures are higher than primary chamber temperatures—at least from 980 to 1,095°C (1,800 to 2,000°F). These values are changeable, sometimes needed to increase, according to the heating value and moisture content of the waste through auxiliary burners located at the entrance to the secondary (upper) chamber to reach at the desired temperatures. This operation enables the demolition of organic toxic substances such as dioxin and furan.

There might be a considerable between the minimum and the maximum waste feed capacities for controlled air incinerators. For example, it ranges from about 0.6 to 50 kg/min (75 to 6,500 lb/hr) at an assumed fuel heating value of 19,700 kJ/kg [8,500 Btu/lb]. On the other hand, lower heating value wastes require a higher degree throughput capacity because feed capacities are limited by primary. The unit size and options purchased are the determining factors on whether waste feed and ash removal will be carried out manually or automatically. [11]

Apart from some advantages, the incineration systems with their various types have some disadvantages. For those negativities, the advanced countries develop some solutions which get rid of the burning smell with flue infiltration at high level costs. If the incinerator

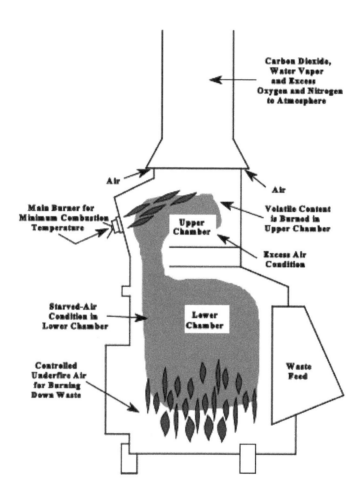

Pic. 12. Incineration Solid Waste Disposal (1995), [22].

operates at lower heat, it produces 8-10 times more dioxin and furans than any normal operation. In other words, dioxins and furans are formed at the maximum level when the incinerator works at the lowest temperature (200-450°C). For this reason, the ambient temperature is increased by using supplemental fuels, therefore, the risk is eliminated. [23]

Pic. 13. 'De Montfort' incinerators

'De Montfort' incinerators are systems widely used by underdeveloped countries today. It is first created by the De Montfort University Innovation Technology Center (Leicester, UK) whose name is now Applied Sustainable Technology Group in order to provide a cheap but effective healthcare wastes incinerators for developing or underdeveloped countries. It would meet the criteria of a temperature of above 800°C with a residence time of over 1 second. [23]

There are considerable amount of studies which show that the implementation of De Montfort incinerators are widely used in certain countries with the low income per capita, such as Kenya, Tanzania, and India. [3]

- Kenya: The studies indicate that the 55% of the 44 De Montfort type incinerators which were constructed in 2002 were in intermittent or regular use. It is also determined that only 1 of 14 sites in which tests and interviews were conducted had an operator with 'near to adequate' skills,[24]. While nearly 40% of health facility managers demonstrated any level of commitment, many technical defects in the equipment and inappropriate operation of incinerators are other results of the observations. [25]
- Tanzania: All 'De Montfort' type incinerators which were constructed in 2001 and 2003 were in use at the time of study of which less than 40% had trained operators, 70% had low smoke disturbance and 60% had safe ash disposal [24]).
- India: HCWH made a research on eight 'De Montfort' incinerators at hospitals in India which are mostly 1 or 2 years old (2002). The findings of this study can be stated as follows: visible smoke from the stack; smoke emission from the chamber door and air inlets; commingling of sharps and non-infectious waste, despite some source segregation; large quantities of unburned materials (sometimes plastics, syringes, glass, paper and gauze) in the ash; deficient ash disposal practices; siting in all cases near populated areas (e.g., playground, orphanage, hospital staff quarters, a primary school, town center) and a lack of operator training. [26]

4.3 Emissions and their monitoring to the incineration technologies

The flue values are really significant for incineration systems. Referring to environmental effects, incinerators can produce toxic emissions such as carbon monoxide (CO), dioxins (polychlorinated dibenzo-para-dioxins or PCDDs), and furans (polychlorinated dibenzofurans or PCDFs).

- Hospital wastes contain a considerable amount of plastic substances. 15-20% of those wastes are PVCs. It is also possible to find approximately 9-10% PVC in infectious wastes. The equipments with PVC which are used in medical centers are IV bags, IV tubes, blood bags, endotracheal tubes, bed coverings, and oxygen tents.

Waste segregation is extremely important for preventing the incineration of PVC (IVs, etc.) waste. In case of any presence of chloric substances such as PVS in the medical waste, this causes the formation of some volatile organic substances as a result of combustion, such as dioxins and furans. The materials with PE or PET are much more preferable than those with PVC.

- Controlling whether the incinerator is constructed in accordance with the recommended dimensions, using appropriate materials, whether it is functioning properly, and keeping the chimney clear of excessive soot.

The incinerators should be constructed in compliance with the standards determined by the relevant authorities of countries. Appropriate material tests and regular flue controls are necessary.

- Ensuring that the preheating of incinerator functions adequately, and that supplementary fuel is added whenever necessary.

Temperature degree should not be ignored. The gases coming out at low temperature constitutes a threat against human health.

- Adopting rigid quality control measures.

It is very important to do routine checks regularly and follow the rules.

In addition to all these instructions, it is also necessary to install a gauge temperature degree on the chimney in order to maintain the temperature during the operation. Smoke levels, loading rates, usage of fuel and the incinerated waste type are other parameters to be rigorously monitored and recorded. If it is impossible to install a gauge, it might be helpful to follow a visual guide in order to maintain the temperature.[26]

Other data to be carefully recorded during the combustion operation: following the temperature of the incinerator, smell degree, the amount of the medical waste which is fed, the amount of the medical waste which is disposed at the end of the day, the amount of the fuel used.

Finally It draws attention towards different systems that the establishment and operation of the incineration technology plants cost very high and the rules and standards are becoming increasingly difficult. The unlimited growth in the residential areas along with the increase in the world population make things difficult for the operators in terms of finding appropriate spaces practically.

The greatest sanction which is impossible to ignore comes from the rule maker authorities. As it can be seen in the publication, namely 'Safe Management of Wastes from Health-care Activities' which was published by WHO in 1999 (Table 4) [8], Dioxins/furans 2.3 ng was declared as TEQ.

In the European Union Directive No. 2000/76/EC [27] which was published on 4 December 2000, the emission values in the incineration plants was determined as 0.1 ng TEQ/m3 for dioxins and furans.

Emission	Daily average (mg/m³)ᵃ	Hourly average (mg/m³)ᵃ	4-hour average (mg/m³)ᵃ
Total dust	5	10	—
Total organic carbon	5	10	—
Chlorine compounds	5	10	—
Fluorine compounds	1	2	—
Sulfur oxides as SO_2	25	50	—
Nitrogen oxides as NO_2	100	200	—
Carbon monoxide	50	100	—
Mercury	—	—	0.05
Cadmium and thallium	—	—	0.05
Lead, chromium, copper, and manganese	—	—	0.5
Nickel and arsenic	—	—	0.5
Antimony, cobalt, vanadium, and tin	—	—	0.5
Dioxins and furans	—	—	0.1
Oxygen content	at least 6% at any moment		

ᵃMeasurements made at standard temperature and pressure.

Table 5. [8] Standards for incinerator emissions in the European Union

As a result of this amendments, the new member states and the Eastern European countries had difficulties with satisfying those criteria, and some of these countries had to close plants, and the others managed to meet those criteria by making unplanned investments. [28] The lifetime for the incineration technologies is limited by 10 years and their establishment and operation is at a high cost. When we make comparison between incineration technologies and non-incineration technologies, the necessity for a considerable maintenance and investment makes non-incineration technologies more preferable.

5. Non-incineration technologies

Non-incineration treatment technologies can be classified based on a number of different criteria. They can be classified according to the ownership in the market share, the types of the waste of disposal, or, the purchase price of the technology. Moreover, Dr. Jorge Emmanuel and his friends' studies in 2004 [28] show another way of categorization non-incineration technologies based on the fundamental process used to decontaminate waste.

Non-incineration Technologies are categorized in this section as follows:

i.   Chemical processes
ii. Irradiative processes
iii. Biological processes
iv. Low-heat thermal processes

5.1 Chemical processes

Chemical disinfectants are generally used for killing microorganisms and inactivating hazardous pathogens. The chemicals used in cleaning the surfaces touched by infected patients and other patients' apparatuses in hospitals is also another method preferred by some systems for nullification of medical wastes. The disinfection process becomes more practical by supplementing a large amount of water into the chemical disinfectants; this also helps to cool mechanical equipment during the shredding, grinding, or mixing process. Those systems that prefers this method are usually used during the volumetric grinding of medical wastes. If the amount of supplemented water exceeds the defined measure, the chemical loses its effect, then it falls short of satisfying the desired level of disinfection. Among the disinfectants preferred in chemical processes are dissolved chlorine dioxide, bleach (sodium hypochlorite), peracetic acid, or dry inorganic chemicals.

Depending on the sterilization effectiveness of the chemical disinfectant to be used in the medical waste disposal, the reaction of some factors should be taken into consideration. Chemical temperature, water and pH values can change the disinfection effectiveness. Chlorine, ozone, formaldehyde, ethylene oxide gas, propylene oxide gas and peracetic acids can be preferred in order to support chemical disinfection. The chemical formulas of some technologies that are available on the market are sold-provided to the user in exchange with kit by the supplier firm of the technology without being exposed.

The most efficient treatment in which the chemical disinfectants are used is of liquid wastes, especially blood, urine, stools, or hospital sewage. On the other hand, there are certain limitations on the chemical disinfection processes of solid- and even highly hazardous- health-care wastes, such as microbiological cultures, sharps, etc. First of all, those types of wastes should be subject to shredding and/or milling processes. It is frequently observed that the shredder, the mostly weakest element of the treatment chain, exposes mechanical failures, or even breakdown. Therefore, the regular check-ups of those equipments are very important. Another point to take into account is the training and the protection of personnel who are supposed to use these powerful disinfectants which are themselves also hazardous. Furthermore, operational conditions should be suitable for an efficient disinfection process. Also, the disinfection should be applied to only the surface of intact solid waste. [8]

5.1.1 Types of chemical disinfectants

Depending on their efficiencies and functions, the type of chemical disinfectants used in the disinfection changes. Sodium hypochlorite functions in killing microorganisms which requires the shredding of the biomedical waste units, such as bags, boxes or other type of containers. For disinfection to be accomplished, there must be a reduction of 4 log 10 (99.99% reduction) in the spores of B. stearothermophilus or B. subtilis. Among other chemicals to be used in chemical disinfection are chlorine derivatives, ozone and some enzymes. [29]

The use of hypochlorite solutions in killing microorganisms, on the other hand, first requires that the biomedical containers be subject to grinding. There must be a reduction of 4 log 10 (99.99% reduction) in the spores of B. stearothermophilus or B. subtilis where the process is completed. For example, chlorine derivatives, ozone or enzymes are some other chemicals that can be used in chemical disinfectant groups.[29]

An overview of the characteristics of the main groups of disinfectants is given in Table 6 [8]

Disinfectants	Bactericidal activity	Tuberculocidal activity	Fungicidal activity	Virucidal activity	Sporicidal activity	Local human toxicity	Applications
Alcohol	Very active	Very active	Very active	Very active	Not active	Moderate	• Skin antisepsis • Disinfection of small surfaces
Chlorhexidine	Less active against Gram-negative bacilli	Not active	Less active	Not active	Not active	Low	• Skin and wound antisepsis
Chlorine compounds (chloramine, hypochlorite)	Very active	Active	Active	Very active	Less active	Moderate	• Skin and wound antisepsis • Water treatment • Surface disinfection
Formaldehyde	Very active	Very active	Very active	Very active	Less active	High	• Disinfection of inanimate objects and surfaces
Glutaraldehyde	Very active	Very active	Very active	Very active	Very active	High	• Disinfection of inanimate objects
Hydrogen peroxide	Less active against staphylococci and enterococci	Active	Active	Active	Less active	Low	• Wound antisepsis
Iodophore	Active	Active	Less active	Active	Not active	Moderate	• Skin and wound antisepsis
Peracetic acid	Very active	Active	Active	Active	Active	High	• Disinfection of inanimate objects
Phenolic compounds	Very active	Very active	Very active	Less active	Not active	High	• Disinfection of inanimate objects and surfaces
Quaternary ammonium compounds	Less active against Gram-negative bacilli	Not active	Less active	Less active	Not active	Low	• In combination with other compounds

Table 6. Characteristics of the main disinfectant groups [8]

As the World Health organization declared and can be seen in the Table 5, there are certain spaces in which disinfectants can be actively used. There is no such a chemical substance which is able to kill all kinds of microorganisms, or suitable for using in every type of medium. As a general fact, it is true that it is the primary function of disinfectants to eliminate microorganisms, or at least reduce them to a certain desired degree. However, their target varies depending on not only their effectiveness but also their corrosiveness. In this regard, it is extremely important to know which type of disinfectant is efficient on which type of medium along with the types of organisms on which they have an effect. Besides the types of chemicals mostly used for disinfection of health-care waste such as aldehydes, chlorine compounds, ammonium salts, and phenolic compounds, ethylene oxide is used in hospitals for the disinfection of the substances of which forms are decomposed at a high degree temperature and heat. Its use is no longer recommended since the handling of ethylene may give rise to significant hazards, as another factor affecting the use of such chemicals in disinfection, although it may still be used in some countries. Under normal circumstances, disposal necessitates single use only. Yet such chemicals are used in some countries intentionally despite their hazards against human health because of their cost-effectiveness. Therefore we can conclude that chemical processes are not sufficient methods in the medical waste disposal by themselves.

5.2 Irradiative processes

Irradiation technology is mainly based on electron beam system. In other words, irradiation technologies comprise Cobalt-60, or ultraviolet beams (UV), or electron beams. This technology works through a process in which electromagnetic radiation extracts the electron from the nuclear orbit when it reaches at a certain energy level. The most significant difference of this technology from microwaves is its extraction of electron from the orbit. The system which completely functions through computer software control is a design of high technology. After the hazardous wastes collected from hospitals are subject to a process through this technology, they become non-noxious- household wastes. The medical wastes become unrecognizable after this processing. They completely lose their physical form and reduce in volume. Considering the advantages of the technology, it is not an unknown system to biomedical engineers since it has been used in cancer therapies more than 20 years. This also shows us that biomedical applications and the biomedical sphere itself are interdisciplinary.

Another advantage of the technology which should be emphasized here is its environment friendly qualities. In contrast to other incineration technologies, it does not cause any air pollution with the minimum degree of toxic emissions except for a negligible amount of ozone. Moreover, the absence of any liquid waste draws another aspect which makes it preferable. In terms of practicality, it is a technology which can be easily operated in room temperature as well as it does not require any extra consumable materials such as extra water, vapor, heat or any extra chemicals. Additionally, it has a low operating cost.

When it comes to the disadvantages of this technology, it requires a highly-organized personnel safety. Since it has a considerable capital cost to build a concrete shield, or an underground structure for the protection from radiation in terms of both thickness and infrastructure, it should be well calculated before the project. Ozone off-gas which is extracted from the exhaust after the process can be considered as another disadvantage of the system.

5.3 Biological processes

In the medical waste disposal technologies, not too many technologies which work on biological procedures are available on the market. The fundamental principle of this technology is to provide the medical waste disposal by destroying enzymes and organic substances. An enzyme mixture used in the system decontaminates the medical waste first, and then the sewage disposal is achieved through the extrusion of the residuum. Only few technologies has tried this method. The system which is not allowed in EU countries has been tested in Virginia State in the U.S.A., under the name "Bio-Converter", and mostly used in agriculture to dispose animal waste due to its capacity for large applications (10 tons/day)[28] Biological process technologies are not widely released to the market because its feasibility are not efficient and these technology still in the research and development phase.

5.4 Low-heat thermal processes

First of all, the equipments designed by the low-heat thermal technologies transform the medical wastes into household (non-noxious) wastes, killing microorganisms in between 93°C and 177°C. We can analyze this technology in two categories: wet heat (steam) thermal technologies and dry heat (hot air) thermal technologies.

- Low-heat thermal technologies:
i.   Autoclaves

In wet heat treatment technology, steam is used, and all thanks to this, the medical waste becomes disinfected and sterilized, that is, disposed of microorganisms. This operation is completed in a pressure container, namely autoclaves. Autoclave technology equipments have been used for years in the sterilization of the materials used in hospitals. These equipments secure the sterilization with pressurized steam at high temperature. Thus, surgery sets in hospitals are being sterilized in this way, and become reusable again and again. The pressurized steam has the capacity to reach at the inmost pores, and thus, to achieve a complete sterilization by disposing of microorganisms.

The medical waste autoclave devices which work through the very same logic kill microorganisms and viruses comprised in the medical waste and make them harmless. The pre-grinding systems available on the market first grind the intact medical waste within a closed system, then inactive the waste of which surface area has expanded, by exposing it to the pressurized steam at a high-degree temperature. The active duration for the pressurized steam to enable sterilization is 15 minutes in 121°C, and 30 minutes in 115°C. When it comes to the time-temperature parameters, the duration of sterilization decreases by the increase in temperature. However, for medical wastes, what is suggested is their nonstop subjection to a 1-4 hour-operation with a minimal temperature of 121°C and a pressure of 2-5 bar (200-500kPa). [29]

The autoclave sterilization clearly exceeds further the expectation that an adequate chemical treatment should be able to inactivate 99.99% of the microorganisms in the waste, that is a level of 4 log 10. On the other hand, the autoclave sterilization provides the killing of microorganisms in all their forms to a level of 6 log 10 or higher. This simply means that a least 99.9999 % of the original spores of *B. stearothermophilus* have been destroyed in the waste, or that only one spore or less has survived the treatment from a population of one million spores. As it can be seen, there is an obvious difference between any chemical treatment and autoclave sterilization.[29]

Advantages of the Autoclaves

As already mentioned, the autoclaves are not a new system, it is a technology that have been already used in hospitals which has been widely accepted in many countries in terms of medical waste treatment. They present such facilities as large availability on the market by virtue of capacity, size and volume.

Although they are a type of hazardous medical waste treatment systems, their emission values are almost non-existing. If exhaust tars outlet pipes are used in the systems, any harm against environment is being prevented by supporting with heat filters. Capital costs are relatively low. This type of technologies that are available on the market is mostly computerized systems, therefore it is possible to see and record the amount of the medical wastes transformed into household wastes.

Disadvantages of the Autoclaves

- Large and sharp entities (shredder or grinder) might be harmful.
- Penetration of the pressurized steam may cause an effective smell in the waste.

The household waste which comes out of the systems without drier or vacuum facilities gains weight because of the pressurized steam to which it has been subjected. The medical waste dumped will be lighter than the household waste gained. The autoclave technologies are not exactly a disposal method but it is a pre-requisite implementation to be done before the medical waste collection and transmission. Any tiny faults which may be done by those who collect and transmit threaten firstly their own health and life, then those of others around them. The autoclave technology which is actively used in hospitals is a factor which helps patient recover after surgery operations by disposing the surgical sets used in those operations of microorganisms on the one hand, it constitutes another factor which might affect community health in a way that if medical wastes are pulled out from hospitals without passing through this system, this may lead any disease onset for some people.

ii.   Microwaves Technology

Microwaves technology simply refers to the inactivation of any hazardous factors in the medical waste with the help of microwaves applied to the water droplets in the waste. Without any external supplementation of water, hot air or pressure, the system works only with the help of microwaves. The microwave frequency to be applied for medical wastes is 2459 MHz, and the wavelength is 12.24 cm.

This system which is commonly preferred in European countries is thought expensive and barely accepted by the Middle East countries. Not as many as those which produce autoclave technologies, there is a number of firms which develop this technology. A certain range of wastes can be treated in a microwave except for volatile and semi-volatile organic compounds, chemotherapeutic wastes, mercury, other hazardous chemical wastes, and radiological wastes.

Advantages of the microwave technology

- Microwave technology has simply no difference with the household microwave ovens, and they are preferable.
- The emissions from microwave units are minimal.
- Even some models on the market do not give out any liquid effluents.
- By using grinders, a volumetric reduction of 80% in the waste can be observed.

There is no difficulty with the use of the technology, and it can be used with one single operator.

The disadvantages of the microwave technology

- There may be some offensive odors around the device.
- The capital cost is relatively high.
- The large and sharp entities (shredder or grinder) may be harmful.

iii.   Dry heat (hot air) Thermal Technologies

Dry heat (hot air) Thermal Technologies present an applicable technology which destroys microorganisms by hot air formed on the surface on which the medical waste is located, using only heat without any pressurized steam and water.

Some technologies are based on the hot wall application in which the waste is heated up by the hot walls of the chamber through conduction and natural convection whereas other technologies use radiant heating by means of infrared and quartz heaters. In comparison to steam-based processes, dry heat processes principally use higher temperatures and shooter exposure times. However, the properties and size of the wastes are determining factors on the time-temperature requisites.[28]

Both systems present similarities with autoclaves and microwaves in terms of the object of the treatment such ascultures and stocks, sharps, materials contaminated with blood and body fluids, isolation and surgery wastes, laboratory wastes (excluding chemical waste), and soft wastes (gauze, bandages, drapes, gowns, bedding, etc.) from patient care. Although the technology allows for human anatomical wastes to be treated, there are ethical, legal, cultural disagreements preventing the use of this technology on these subjects.

In certain systems with exhaust outlet pipes, it is possible to filter by using a combination of HEPA, carbon filters and venture scrubber which reduces offensive odors to a minimal level.

Advantages of the dry heat thermal technology

- Its structure is not too complicated in terms of design.
- It is a reliable system to inactivate wet and liquid wastes (body fluid, blood, etc.).
- There is no liquid effluents.
- There is a volumetric reduction of 80% in the models with internal shredders.
- It can be used by one single operator.

Disadvantages of the dry heat thermal technology

- Large and sharp entities (shredder or grinder) might be harmful.
- It may not be preferable for hospitals due to considerable power expenditure.
- There may be some offensive odors around the device.

In Table 7 some of non-incineration technologies listed. For further information, detail can see reference list.

Non-incineration Technology	Technology name, vendor
LOW-HEAT THERMAL PROCESSES	
Autoclave or Retort	Tuttnauer
Shredding-Steam-Mixing/Drying	Ecodas
Steam-Mixing-Fragmenting/Drying	Hydroclave Systems Corp.
Shredding-Steam-Mixing/Drying, Chemical	Steriflash, T.E.M.
Vacuum-Steam/Drying/Shredding	Sterival, Starifant Vetriebs GmbH
Shredding-Steam/Drying, Chemical	STI Chem-Clav, Waste Reduction Europe Ltd
Shredding-Steam-Mixing/Compaction	STS, Erdwich Zerkleinerungssysteme GmbH

Non-incineration technology	Technology name, vendor
Vacuum-Steam/Drying/Shredding	System Drauschke, GÖK Consulting AG
Steam-Fragmenting/Drying	ZDA-M3, Maschinenvertrieb für Umwelttechnik GmbH
Microwave Treatment	Ecostéryl, AMB S.A.
Microwave Treatment	Medister, Meteka
Microwave Treatment	Sanitec
Microwave Treatment	Sintion, CMB Maschinenbau und Handels GmbH
Vacuum-Steam/Microwave Treatment/Drying/Shredding	Sterifant Vertriebs GmbH
CHEMICAL PROCESSES	
Fragmenting-Steam-NaClO/Cl_2O	Newster, Multiservice First s.r.l.
Alkaline Hydrolysis	WR^2, Waste Reduction Europe Ltd
IRRADIATION PROCESSES	
Electron Beam-Shredding	U. Miami E-Beam
BIOLOGICAL PROCESSES	
-	Today not available in Europe

Table 7. Some of non-incineration technologies listed. For further information see Reference [28], [29]

6. Conclusion

When Medical Waste Treatment Technologies are presented as an example for Biomedical Instrument Application, it has been revealed that the technologies which can be used in different areas are also applicable for different purpose. Therefore they may have direct or indirect effects on human health.

For example, while the autoclave technologies that have been used in hospitals for years are systems which help patients recover by disposing the surgical sets used in operations of microorganisms (if this system is used correctly, it efficiently reduces the risk of infection for the patient), they are necessary to be applied during the medical waste collection and transmission in order to eliminate all possible risks although they are not exactly disposal methods in low-heat thermal technologies. Any tiny faults which may be done by those who collect and transmit threaten firstly their own health and life, then those of others around them. While hospitals are certain places for some people to find their remedies, they should not become a kind of place which causes some people to get diseases.

As rules and laws vary depending on countries, this chapter is written in order to give a presentation of and raise an awareness on the operations in Asia, Europe and America. For example, while there is no permission for medical waste storage in European countries, hospitals in Asia and non-member of EU are storing their medical wastes in those hospitals for weeks and even months, or leave them on the streets without protection and lead to a danger against environment. Street animals become agents of an unhealthy cycle by carrying those microbes to humans. Unconscious Landfilling operated in some countries, as shown, has a negative effect on human life. The increase in the world population along with the carelessness in taking necessary precautions on time makes the subject more crucial, even it has been tried to attract attention to the awareness of the significance of the problem by some social organizations.

While the developed countries pride themselves on the medical waste statistics of their hospitals, a low level of medical waste extracted can be considered as a matter of success in some countries. The developing technologies and the investments of health care sector indicate that the care for human health is only possible to use different materials for each patient and prevent possible risks rather than using the same materials again and again. The infection and medical waste statistics among hospitals should be declared by the relevant authorities, a competence between hospitals should be raised.

The statistics to be published will show that people will prefer the hospitals with low infection and high medical waste statistics in the countries in which they live.

As an interdisciplinary area, biomedical has taught us the methods for considering different matters from different aspects. We should expose what we can do, both technologically and instrumentally, on the matters of which we have a certain level of true knowledge and thoughts. We should know how to sustain our lives without harming the world in which we live, without putting at risk others' lives. We should aim at leaving a living space for our future, for our children which will be better than what we have had.

For further information, primary references used in this chapter of the book are recommended: Safe Management Wastes from Health-care Activities co-edited by A. Prüss, E. Giroult, P. Rushbrook which was published by World Health Organization in GENEVA in 1999; and "Non--Incineration Medical Waste Treatment Technologies: A Resource for Hospital Administrators, Facility Managers, Health Care Professionals, Environmental Advocates, and Community Members" prepared by Dr. Jorge Emmanuel in 2004 which was issued by Health Care Without Harm.

7. References

[1] The Ministry of Environment of Turkish Republic, (22 June 2005/25883). "The Regulation for Medical Waste Control" available at *www.cygm.gov.tr/CYGM/Files/mevzuat/yonetmelik/tibbi.doc* Accessed 01 January, 2012.

[2] World Health Organization (WHO), (November 2011) available at http://www.who.int/mediacentre/factsheets/fs253/en/ Accessed 12 January, 2012.

[3] Batterman S. (2004). Findings on an Assessment of Small-scale Incinerators for Health-care Waste. WHO, Geneva, Switzerland.

[4] Directive 75/442/EC and transposed in order to obtain a cognitive national and European data on waste. (15.07.1975)

[5] Directive 2006/12/ECC Of The European Parliament And Of The Council of 5 April 2006 on waste (O.J. L 114, 27.04.2006)

[6] The Commission Resolution, Decision 2000/532/EC of 3 May 2000 replacing Decision 94/3/EC establishing a list of wastes pursuant to Article 1(a) of Council Directive 75/442/EEC on waste and Council Decision 94/904/EC establishing a list of hazardous waste pursuant to Article 1(4) of Council Directive 91/689/EEC on hazardous waste (O.J. L 226, 06.09.2000).

[7] Council Directive 91/689/EEC of 12 December 1991 on hazardous waste (O. J. L 377 , 31.12.1991).

[8] WHO Safe management of wastes from health-care activities co-edited by A. Prüss, E. Giroult, P. Rushbrook which was published by World Health Organization in GENEVA in 1999

[9] Baldwin, CL; Runkle, RS, (1967) Biohazards Symbol: Development of a Biological Hazards Warning Signal, volume 158 (issue 3798): page 264-265

[10] European Union Council Directive 67/548/EEC of 27 June 1967 on the approximation of laws, regulations and administrative provisions relating to the classification, packaging and labelling of dangerous substances

[11] The Ministry of Environment of Turkish Republic, (22 June 2005), "The Regulation for Medical Waste Control 25883" the Appendix 2

[12] Official Journal of the European Communities (16 January 2001), (2001/118/EC), available at http://eur-lex.europa.eu/LexUriServ/LexUriServ.do?uri=OJ:L:2001:047:0001:0031 :EN:PDF Accessed 11 September, 2011.

[13] AODA, (America's Ocean Dumping Act) picture available at ; http://www.greenwala.com/channels/nature/blog/5741-The-Ocean-s-Top-25-Deadliest-Pollution-Predators Accessed 12 October, 2011.

[14] WHO | Waste from health-care activities, Chapter 3: "Safe management of wastes from health-care activities" Table 3.2 Viral hepatitis B infections caused by occupational injuries from sharps, Available at www.who.int/entity/water.../020to030.pdf

[15] Ozturk M. (2009) *Medical Waste Management in the Hazardous Waste Group, the Report of the Deputy Chair of the Environment Commission of the Grand National Assembly of*

Turkey Medical Waste Management in the Hazardous Waste Group, the Report of the Deputy Chair of the Environment Commission of the Grand National Assembly of Turkey

[16] Princeton University (2008), "Laboratory Waste Streams", Picture. 5. available at; http://web.princeton.edu/sites/ehs/biosafety/livevirusworker/decontamination. htm Accessed 10 Jan, 2012.

[17] Hazardous Waste Management Plan (HWMP), (1990), California Country of El Dorado, available at; http://www.co.eldorado.ca.us/Government/EMD/HazardousMaterials/Medical _Waste.aspx

[18] UNEP (United Nations Environment Programme environment for development), (22 Jun 2007), "Waste pickers at the main Khartoum landfill site; Waste management is problematic throughout Sudan", available at http://postconflict.unep.ch/sudanreport/sudan_website/index_photos_2.php?ke y=waste%20management Accessed 1 Jan 2012.

[19] Council Directive 1999/31/EC of 26 April 1999 on the landfill of waste entered into force on 16.07.1999. The deadline for implementation of the legislation in the Member States was 16.07.2001.

[20] Oeltzschner H, Mutz D (1996). *Guidelines for an appropriate management of domestic sanitary landfills*. Eschborn, Gesellschaft für Technische Zusammenarbeit.

[21] Batterman S, (Website Updated April 9, 2008), Mozambique Program for Healthcare facility waste treatment, The Regents of the University of Michigan School of Public Health (MSPH), available at http://research.sph.umich.edu/project.cfm?deptID=2&groupID=1&projectID=18 Accessed 12 September, 2011.

[22] U.S. Environmental Protection Agency, (1995) "Emission Factor Documentation for AP-42 SECTION 2.6 MEDICAL WASTE INCINERATION" North Carolina, available at http://www.epa.gov/ttnchie1/ap42/ch02/bgdocs/b02s03.pdf Accessed 12 April, 2012.

[23] Picken D. J. (2007). De Montfort medical waste incinerators. Available at http://www.mwincinerator.info/en/101_welcome.html Accessed 8 November, 2011.

[24] Adama S. (2003). Results of Waste Management in Africa 2003. Presentation at the Safari Park Hotel, Nairobi, Kenya, 18-20 September 2003

[25] Taylor E. (2003). Rapid Assessment of Small-scale Incinerators: Kenya, Final Draft

[26] Veronica D. B.(2002), Practical Action Technology Challenging poverty,("Health-Care Waste Management in Developing Countries" available at http://practicalaction.org/low-cost-medical-waste-incinerator   Accessed   12 December, 2011.

[27] Directive 2000/76/EC of the European Parliament and of the Council of 4 December 2000 on the incineration of waste. Available at http://www.central2013.eu/fileadmin/user_upload/Downloads/Document_Cent re/OP_Resources/Incineration_Directive_2000_76.pdf

[28] Emmanuel J. (June 2004) Non-Incineration Medical Waste Treatment Technologies HCWH

[29] Ontario, (2002) "Non-Incineration Technologies for Treatment of Biomedical Waste" (Procedures for Microbiological Testing), Sections 19 and 27; Part XVII, Section 197. GUIDELINE C-17

5

Biosensors and Their Principles

Ahmet Koyun[1], Esma Ahlatcıoğlu[1] and Yeliz Koca İpek[2]
[1]Yıldız Technical University, Science and Technology Application and Research Center,
[2]Tunceli University, Faculty of Engineering, Department of Chemical Engineering,
Turkey

1. Introduction

Biological and biochemical processes have a very important role on medicine, biology and biotechnology. However, it is very difficult to convert directly biological data to electrical signal, the biosensors can convert these signals and the biosensors over this diffuculty. In recent years, thanks to improved techniques and devices, the usage of these products have increased.

The first biosensor was described in 1962 by Clark and Lyons who immobilized glucose oxidase (GOD) on an amperometric oxygen electrode surface semipermeable dialysis membrane in order to quantify glucose concentration in a sample directly [1, 2]. They described how "to make electrochemical sensors (pH, polarographic, potentiometric or conductometric) more intelligent" by adding "enzyme transducers as membrane enclosed sandwiches".

According to a recently proposed IUPAC definition [3], " A biosensor is a self-contained integrated device which is capable of providing specific quantitative or semi-quantitative analytical information using a biological recognition element (biochemical receptor) which is in direct spatial contact with a transducer element. A biosensor should be clearly distinguished from a bioanalytical system, which requires additional processing steps, such as reagent addition. Furthermore, a biosensor should be distinguished from a bioprobe which is either disposable after one measurement, i.e. single use, or unable to continuously monitor the analyte concentration".

A biosensor is a device composed of two elements:

1. A bioreceptor that is an immobilized sensitive biological element (e.g. enzyme, DNA probe, antibody) recognizing the analyte (e.g. enzyme substrate, complementary DNA, antigen). Although antibodies and oligonucleotides are widely employed, enzymes are by far the most commonly used biosensing elements in biosensors.
2. A transducer is used to convert (bio)chemical signal resulting from the interaction of the analyte with the bioreceptor into an electronic one. The intensity of generated signal is directly or inversely proportional to the analyte concentration. Electrochemical transducers are often used to develop biosensors. These systems offer some advantages such as low cost, simple design or small dimensions. Biosensors can also be based on gravimetric, calorimetric or optical detection [1].

Biosensors are categorized according to the basic principles of signal transduction and biorecognition elements. According to the transducing elements, biosensors can be classified as electrochemical, optical, piezoelectric, and thermal sensors [3]. Electrochemical biosensors are also classified as potentiometric, amperometric and conductometric sensors.

The application of biosensor areas [4] are clinic, diagnostic, medical applications, process control, bioreactors, quality control, agriculture and veterinary medicine, bacterial and viral diagnostic, drag production, control of industrial waste water, mining, military defense industry [5], etc. A few advantages of biosensors are listed below:

1. They can measure nonpolar molecules that do not respond to most measurement devices
2. Biosensors are specific due to the immobilized system used in them
3. Rapid and continuous control is possible with biosensors
4. Response time is short (typically less than a minute) and
5. Practical

There are also some disadvantages of biosensors:

1. Heat sterilization is not possible because of denaturaziation of biological material,
2. Stability of biological material (such as enzyme, cell, antibody, tissue, etc.), depends on the natural properties of the molecule that can be denaturalized under environmental conditions (pH, temperature or ions)
3. The cells in the biosensor can become intoxicated by other molecules that are capable of diffusing through the membrane.

2. Recent development topics on biosensors

In biosensor development studies, suitable bioreceptor molecule, suitable immobilization method and transducer should be selected firstly. Biology, biochemistry, chemistry, electrochemistry, physics, kinetics and mass transfer knowledge is reuired for this study. Thus we can say that developing a biosensor is related with a interdisciplinary study. Proportional to the technological development and increase of interdisciplinary studies biosensors are being more useful and having more usage areas day by day. Recent development topics which are listed below will be discussed in this chapter:

• Electrochemical biosensor
• Fiber-optic biosensor
• Carbon Nanotube
• Protein Engineering for biosensors
• Wireless Biosensors Networks

2.1 Electrochemical biosensors

Bioelectroanalysis with electrochemical biosensors is a new area in rapid development within electroanalysis. In biosensor development studies, suitable bioreceptor molecule, suitable immobilization method and transducer should be selected firstly Bioelectroanalytical sensors permit the analysis of species with great Specificity, very rapid, sensitive, highly selective and cheap cost in principle. They can be used in clinical analysis, in on-line control processes for industry or environment, or even in vivo studies [6]. The difference between biosensor and physical or chemical sensors is that its recognition element is biological.

The investigated bioelectrochemical reaction would generate a measurable current (amperometric detection), a measurable potential or charge accumulation (potentiometric detection) or measurable conductivity change of a medium (conductometric detection) between electrodes. When the current is measured at a constant potential this is referred to as amperometry. If an electrical current is measured while controlled variations of the potential is being applied, this is named as voltammetry.

Potentiometric, amperometric and conductometric measurement techniques forms the kinds of electrochemical biosensors. Potentiometric sensors have an organic membrane or surface that is sensitive to an analyte. The reaction between them generates a potential (emf) proportional to the logarithm of the electrochemically active material concentration. This potential is compared with the reference electrode potential.

Enzyme immobilized electrodes reacts with substrate and products are detected by electrodes. Amperometric sensors measure the current change resulted by chemical reaction of electroactive materials while a constant potential is being applied. The change of the current is related to the concentration of the species in solution.

Generally biological compounds (glucose, urea, cholesterol, etc.) are not electroactive, so the combination of reactions to produce an electroactive element is needed. This electroactive element leads a change of current intensity. This change is proportional to the concentration of analyte.

Conductometric biosensors can measure the change of the electrical conductivity of cell solution. Most reactions involve a change in the composition of solution. Thus conductometric biosensors can detect any reactive change occuring in a solution.

Electrochemical biosensors have advantages that they can sense materials without damaging the system [7]. The use of biosensors for industrial and environmental analysis [8] is very important. The control of food manufacturing processes, evaluation of food quality, control of fermentation processes and for monitoring of organic pollutants are some of the applications of biosensors. The present popularity of analytical biosensors is due to their specific detection, simple use and low cost. For example an electrochemical biosensor can be used to detect Salmonella and E. coli O157:H7 in less than 90 min. [7]. Electrochemical biosensor studies are performed with electrochemical cells.

Electrochemical Cells

An electrochemical cell is used in electrochemical sensor studies. The electrodes themselves play an important role in the performance of electrochemical biosensors. The electrode material, its surface modification or its dimensions effects the detection ability of the electrochemical biosensor. There are three kinds of electrodes in the electrochemical cell:

- Working electrode
- Reference electrode
- Auxilary (counter) electrode

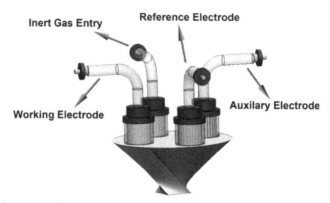

Fig. 1. Electrochemical cell.

Reference electrode:

The other electrodes in the cell are referred to this electrode. Reference electrode types:

- Type 1: the hydrogen electrode
- Type 2: the calomel electrode
- Type 3: glass electrodes

Reference electrode is a kind of standard hydrogen electrode. Hydrogen is potentially explosive and is not very suitable using an electrode with hydrogen gas for routine measurements. So there are two common use and commercially available reference electrode types:

- Ag/AgCl Electrode: There is a Ag wire that coated with AgCl and dipped into NaCl solution.

$$AgCl + e- \rightarrow Ag + Cl^- \qquad (E^0 = +0.22V)$$

- Saturated-Calomel Electrode: Calomel is the other name of mercurous chloride (Hg_2Cl_2).

$$Hg_2Cl^2 + 2e- \rightarrow 2Hg + 2Cl^2 \qquad (E^0 = +0.24V)$$

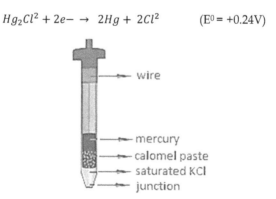

Fig. 2. Reference (calomel) electrode.

Calomel electrode is consist of mercury, paste (mixture of mercury(I) chloride powder and potassium chloride) and saturated potassium chloride solution.

Auxilary (Counter) Electrode:

In a two-electrode system, when a known current or potential is applied between the working and auxiliary electrodes, the other variables may be measured. The auxiliary electrode functions as a cathode whenever the working electrode is operating as an anode and vice versa. The auxiliary electrode often has a surface area much larger than that of the working electrode. The half-reaction occurring at the auxiliary electrode should occur fast enough not to limit the process at the working electrode. The potential of the auxiliary electrode is not measured against the reference electrode but adjusted to balance the reaction occurring at the working electrode. This configuration allows the potential of the working electrode to be measured against a known reference electrode. Auxiliary electrode is often fabricated from electrochemically inert materials such as gold, platinum or carbon.

Working Electrode:

It is the electrode on which the reaction occur in an electrochemical system [9, 10, 11]. In an electrochemical system with three electrodes, the working electrode can be referred as either cathodic or anodic depending on the reaction on the working electrode is a reduction or an oxidation. There are many kind of working electrodes. Glassy carbon electrode, screen printed electrode, Pt electrode, gold electrode, silver electrode, Indium Tin Oxide coated glass electrode, carbon paste electrode, carbon nanotube paste electrode etc.

Screen printed electrodes are prepared with depositing inks on the electrode substrate (glass, plastic or ceramic) in the form of thin films. Different inks can be used to get different dimensions and shapes of biosensors. Screen-printed electrochemical cells are widely used for developing amperometric biosensors because these biosensors are cheap and can be produced at large scales. This could be potentially used as disposable sensor that decreases the chances of contamination and prevents loss of sensitivity. Figure 3. exhibits an electrochemical biosensor as screen printed electrode.

Performance factors of an electrochemical biosensor are: Selectivity, response time, sensitivity range, accuracy, recovery time, solution conditions and the life time of the sensor.

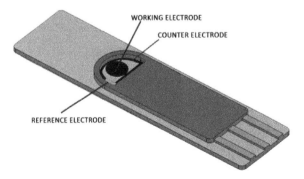

Fig. 3. Electrochemical biosensor as screen printed electrode.

Cyclic voltammetry or CV

Cyclic voltammetry is a type of potentiodynamic electrochemical measurement. In a cyclic voltammetry experiment, the working electrode potential is changed linearly versus time. Cyclic voltammetry experiment ends when it reaches a set potential value. When cyclic voltammetry reaches the set potential, potential ramp of the working electrode is inverted back. This inversion can happen multiple times during a single experiment until a set cycle number is obtained. The plot of the current at the working electrode vs. the applied voltage give the cyclic voltammogram of the reaction. Cyclic voltammetry is a general way to study the electrochemical properties of an analyte in a solution [12, 13, 14].

Chronoamperometry and Chronopotentiometry

A potential is applied to the working electrode and steady state current is measured as a function of time for chronoamperometric measurement. There is a diffusion layer between solution media and electrode surface. The concept of a diffusion layer was introduced by Nernst. Diffusion controls the transfer of analyte from the bulk solution of higher concentration to the electrode. Thus there is a concentration gradient from solution media to the electrode surface. Cottrell equation can indicate this situation better: It defines the current-time dependence for linear diffusion control at an electrode.

$$I = nFAc_0 \sqrt{\frac{D}{\pi t}}$$

I: current is dependent on
F: Faraday's constant,
n: The number of transferred electrons for each molecule,
A: The electrode area,
c_0: The analyte concentration,
D: The diffusion coefficient and time
t: Time

Electrochemical Impedance Spectroscopy (EIS)

Electrical resistance can be described as the ability of a circuit element to resist the flow of electrical current. This is defined with Ohm's law:

E=IxR for DC conditions

While this is a well known equation, its use is limited to only the ideal resistor. An ideal resistor follows Ohm's Law at all current and voltage levels and its resistance value is independent of frequency.

Impedance is a measure of the ability of a circuit to resist the flow of electrical current Like resistance, but electrochemical impedance is usually used by applying an AC potential to an electrochemical cell and then measuring the current through the cell. When we apply a sinusoidal potential, the response to this potential is an AC current signal.

This current signal can be considered as a sum of sinusoidal functions (a Fourier serie). For AC conditions: E = IxZ, where Z is the impedance of the system. The impedance can be calculated by setting the input potential and measuring the induced current.

Electrochemical impedance spectroscopy (EIS) is a technique well suited for evaluating coating permeability or barrier properties for corrosion control of steel structures based on the electrical resistance of the coating. EIS has been widely used in the lab to determine coating performance and to obtain quantitative kinetic and mechanistic information on coating deterioration [15].

Detection of Analyte

Detection principle of analytes changes according to transducer type of the biosensor. Electrochemical biosensors use electrical signals as output datas. Thus detection of an analyte is related with the changes of electrical signals. For example; the intensity of the current, potential energy and electrical conductivity of the electrode change.

In cyclic voltametry studies, It is seen that scientists observe the electrical potential vs. electrode current intensity of an electrochemical cell system. When the analyte reacts with a biological component that coated or immobilized on the electrode surface, a change in electrical current occur at an electrical potential array. This current change tells us that there is an electron transfer in the electrochemical cell during the reaction between the analyte and biological component of biosensor electrode. In Figure 4 an example of electrochemical biosensor study for monocrotophos detection with acetyl choline esterase (AChE) enzyme immobilized on a modified glassy carbon electrode (GCE) with Au NanoParticles-SiSG is given below.

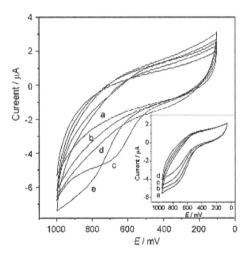

Fig. 4. Cyclic voltammograms of GCE (a) and AChE-AuNPs-SiSG/GCE (b) in pH 7.0 PBS; AChE- AuNPs -SiSG/GCE (c), AuNPs -SiSG/GCE (d) and AChE-SiSG/GCE (e) in pH 7.0 PBS containing 1.0 mM ATCl. Inset: Cyclic voltammograms of AChE-AuNPs-SiSG/GCE in pH 7.0 PBS containing 1. 0mM ATCl after immersed in 0 (a), 0.01 (b), 0.2 (c) and 5 (d)_g/ml monocrotophos solution, respectively, for 10 min [16].

The current intensity diffecence between a and d in inset CV graph gives the result for monocrotophos concentration (the analyte). The inhibition of the enzyme is used for the detection of analyte in this example.

Immobilization methods

Electrochemical detection techniques use predominant enzymes. Because enzymes have specific binding capabilities and biocatalytic activity. Some of the other biorecognition elements are antibodies, nucleic acids, cells and micro-organisms. Biorecognition elements should be immobilized on the electrode surface. Adsorption, microencapsulation, entrapment, covalent attachment and cross linking methods Are the most well known immobilization methods.

Adsorption methods:

1. Physical adsorption (physisorption) and
2. Chemical adsorption (chemisorption).

Physisorption is weaker than chemisorption. Adsorption is the simplest way for immobilization of organic material, however the bonding is weak and life time of electrode is short.

Microencapsulation method is more reliable for adsorption. In this method, an inert membrane traps the biologic material on the working electrode. Most used membranes are cellulose acetate, collagen, gluter aldehyde, chitosan, nafion, polyurethanes, etc..

In entrapment method, generally a solution of polymeric materials are prepared containing biologic material that will be entrapped onto the working electrode. The solution is coated on the electrode with various coating methods. Starch gels, nylon and conductive polymers such as polyaniline or nafion are used for.

Covalent attachment immobilization is important particularly for the advantage that the enzyme is not been released from the electrode surface when it is used. However, covalent bonding should not decompose or hide the active site of the enzyme. The functional groups that may take part in this binding are NH_2, CO_2H, OH, C_6H_4OH and SH groups. [15].

Cross linking is bonding two or more molecules by covalent bonds. In cross-linking method bifunctional agents such as gluteraldehyde are used to bind the biological materials. The disadvantage of this method is high ratio of enzyme activity loss.

2.2 Fiber optic biosensor

The optical fiber is flexible and has small wires generally made out of glass or plastic in different configuration, shape, and size. It can transmit light signals for long distances with minimum lost value. The optical fiber is convenient for harsh and hazardous environments, because of their remarkably strong, flexible and durable structures. It is non-electrical; therefore, it can be used in various damaged electric current applications. Optical fibers are commonly used because of high quality and its low cost for sensing applications. Particularly, the main attractive properties of optical fibers can permit transmission of multiple signals synchronously and by this means it can obtain multiple capabilities for sensing of analyte [17].

Figure 5 exhibits the optic fibers model that is containing a core and coating. Their refractive indices are n1 and n2, respectively as shown in Figure 6. The core and cladding interference act as mirror because of their different refractive indices (Fig 7a) [17].

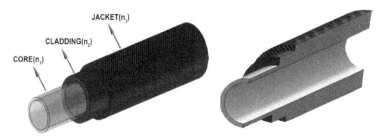

Fig. 5. Optic Fiber

The core and cladding parts play a very important role particularly on the light transmission. Their refractive indices are n_1 and n_2, respectively (Fig 5). In the Fig 6a the core and cladding interference act as mirror because of their different refractive indices. The series of internal reflections transmit the light from one end of the fiber to the other one [17].

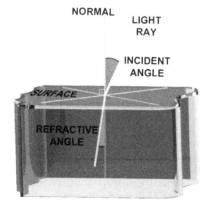

Fig. 6. TIR Principle

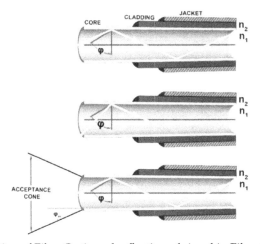

Fig. 7. The lateral section of Fiber Optic and reflection of signal in Fiber Optic

Generally fiber optic biosensors (FOB) work with using total internal reflection (TIR) principle (Fig 6) if two main conditions are satisfied:

i= Compared with the critical angle, larger cladding angle facilitate the reflection of the light and spread it through the fiber.

θ_C= Critical angle

n_0= Refractive index of medium
n_1= Refractive index of core
n_2= Refractive index of cladding

$$\theta_C = n_2/n_1$$

i.   Light angles entering through the fiber should be within the acceptance cone as shown Figure 7c. The acceptance cone angle, θ_m depends on refractive indexes of core, clad and medium.

$$\sin\emptyset_m = (n_{12}-n_{22})n_0$$

$$\sin\emptyset_m = \frac{\sqrt{(n_1{}^2-n_2{}^2)}}{n_0}$$

Another parameter is numerical aperture. The relation between numerical aperture and acceptance cone's angle is shown as follow equation:

$$NA = n_0 \sin\emptyset_m$$

The light collecting capabilities of the fiber is high when the acceptance cone is wide. The larger the NA is, the more powerful optic fiber will be [17].

Fiber Optic is used in optical fiber biosensors that measure some biomolecules such as proteins, nucleic acids etc.) Because of the attractive properties of fiber optic biosensor such as low cost, efficiency, accuracy, these take place of literature and they are preferred in many applications.

The Fiber Optic biosensor provides alternative measurements method to conventional methods for determination of biological species.

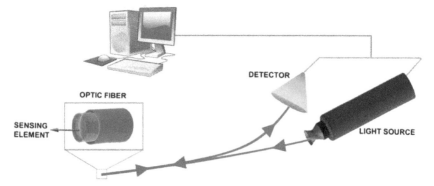

Fig. 8. The Fiber Optic Biosensor

The basic system of a fiber optic biosensor consists of a light source, an optical fiber, sensing material and a detector. An optical fiber transmits the light and also acts as the substrate for the sensing material. Detector measures the output signal. (Fig8) Some light source of optical biosensors are tungsten lamp, deuterium lamp, xenon lamp, LEDs, Laser, Laser diodes and some light detectors for optic biosensors are avalanche photodiodes, photodiodes, photomultipliers, charge- coupled devices [17].

When the reaction occurs between sensing element and the analyte, there is a change both its physico-chemical and optical properties. This transduction mechanism, generates optical signals, is related with analyte concentration. To measure the optical signals, the difference between incident and output light is determined at the location where the sensing element is fixed. Output light is send to detector by fiber. Collected light (reflected, emitted, absorbed light) is measured on the detector. [17].

The Fiber Optic Biosensor have some advantages and disadvantages which are shown below.

The Advantages of Fiber Optic Biosensor [18, 19]

1. There is no need reference electrode in the system
2. It can be easily moved, because there is no reagent in contact of any optical fiber
3. There are no electrical safety hazards and electrical interference
4. It is less dependent than temperature compared with electrode
5. It can be found in-vivo measurement applications because of easy miniaturization
6. Multiple analytes can be determined thanks to guide the light in different wavelengths at the same time.
7. It can be used for the most of chemical analytes because of its spectroscopic properties.

The Disadvantages of Fiber Optic Biosensor

1. The life time of the reagents can be short under incident light
2. Because of the diffusion of analytes, it may cause slow response time
3. Fiber Optic Biosensor only works for spesific reagent.
4. Optimized commercial accessories have limited availability when using them with optical fibers.

The Types of Fiber Optic Biosensors

Absorbance Fiber Optic Biosensor: An atom or a molecule absorbs light energy is called as absorption. The molecule takes this energy and moves to higher excited energy state from ground energy state.

Lambert Beer Law is used for the absorption.

$$A = \log\left(I_0/I\right) = \varepsilon.[C].l$$

A=Optical absorbance
I_0= incident light intensity
I= transmitted light intensity
l=effective path length
ε=Molar absorption coefficient

Fig. 9. The types of fiber optic biosensors

Practically, the optical fibers detect the transmitted and scattered light through the fiber and then it can be obtained absorbance values.

Fluorescence Fiber Optic Biosensor: Fluorescence is commonly used in fiber optic biosensors and better than adapted by optical sensors compared to absorption fiber optic biosensors and the other advantage is very sensitive technique that can detect very low concentrations [20].

When the molecule excited, they gain some energy to move to higher energy state which is non-stable state. After that they want to return the ground state because of conservation their steady state. In fluorescence optic fiber biosensors, fluorescence signals are measured by transmitting the excitation light through an optical fiber and the light emission is measured via detector. Generally, it is measured using the change of fluorescence intensity and related to the analyte concentration [17].

Luminescence Fiber Optic Biosensor: Luminescence can be mainly classified by two parts. These are chemiluminescence and bioluminescence. On the contrary to fluorescence, excited species are obtained as yield of chemical reaction and these excited species emit light while returning to the ground state. Aboul- Enein et al. studied chemiluminescence in fiber optic biosensors [21]. The bioluminescence, a biological chemi-luminescent reaction, is produced by many living organisms in nature for mating, self-protection and finding food [22].

As a simple example, if a wide diversity of sequence of biochemical reactions is used, the production of light will increase. This enzymatic reaction is catalyzed by luciferase and liberates a compound in its excited state while it is going back to its ground state.

The mechanism of light emission of Oxyluciferin* is similar to fluorescence that can be produced by irradiating oxyluciferin via the standard method

Reflectance Fiber Optic Biosensor: The reflectance fiber optic biosensor works with evanescence waves. Besides transmittance and absorbance, reflectance of analyte is another measurement method. The reason of reflectance changing is the structure of material. The effect of bio-interface reflectance changes in a large band.

In recent years, fiber optic biosensors has been very useful for the medical technology, dramatically improving patient care and cutting overall operating costs. Nowadays, they are currently used in a variety of medical application such as early cancer and AIDS detection.

2.3 Carbon nanotube biosensor

The most of the scientists have claimed that a coupling of material science and biology in the nanosize will have a remarkable effect on the many fields of science and technology. Particularly in the biology field, nanosize is very important scale because many important biomacromolecules structures are in the range of 1-1000nm. [23]

Because of these reasons, the focus is on nanostructured materials. It helped develop the unique properties of new devices and sensors. These nanostructured materials have good chemical sensitivity, biocompatibility, and good electrical sensitivity with changes of chemical composition. The sensitive materials have played a significant role for the chemical and biological sensor because of their sizes which are close to biomolecules.

The performance and improvement of biosensors highly depend on the materials. Moreover the chosen materials of making transducer are directly related to their physicochemical characteristics.

The carbon materials such as carbon nanotubes (CNTs) are used in making biosensor.

CNT's are well ordered and hexagonal arrangements of Carbon atoms which have been rolled into tubes. It can be considered as the cylindrical graphite layer or layers which have nanometer scale of diameter. Therefore, it can be classified as single wall (SWNT) and multiwall carbon nanotube (MWNT) as structural. The diameter of SWNT is approximately 0.4-2 nm and the other one is 2-100 nm.

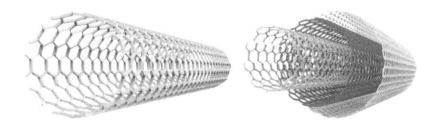

Fig. 10. Single and Multi- walled Carbon Nanotubes, respectively.

They can also be metallic conducting or semiconducting carbon nanotubes which change with geometrical structure. The chiral angle, which determines the twisting value of CNT, play important role on the conductivity of CNTs. It can be called as zig zag, armchair and chiral structure (Fig. 12)

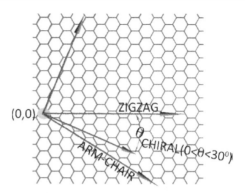

Fig. 11. The unrolled, two-dimensional, honeycomb lattice of a CNT

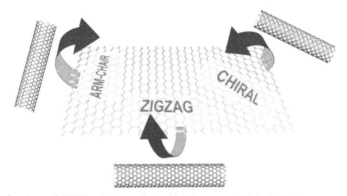

Fig. 12. Classification of CNTs, a) Arm-chair, b) Zig-zag, c) Chiral CNTs

In particular, it was explained the relation with structure and electrical conductivity of SWNTs by some researchers. The studies have shown that arm-chair tubes can be metallic. Beside this, zig-zag and chiral tubes can be either metallic or semiconducting. The conductivity values are related with the wrapping angle and the length of CNTs [24].

The electrical measurements of MWCNTs have shown that the electrical conductivity of MWNTs can be metallic or semiconducting character [24, 25].

Major methods of CNT synthesis are electrical arch discharge, laser vaporization, and chemical vapor deposition (CVD) [26, 27, 28, 29, 30, 31].

CNTs exhibit attracted electrocatalytic activity because of their interesting properties such as their dimension, electronic structure etc [26, 28, 32, 33-35].

Generally, in the voltammetric response of several molecules at electrodes modified with CNTs, higher peak currents and lower overvoltage are observed. Reading the literature, it can be claimed that CNTs is a very challenging materials for the preparation of electrochemical sensors due to these unique properties [26, 36-43].

They have some advantages such as small size, high strength (approximately 100 times higher than the strength of steel), high electrical (approximately 100 times greater than for cupper wires) and thermal conductivity (higher than diamond), high specific surface area, simple preparation, less power, long term stability, good reproducibility, fast response etc. Therefore, CNT has better properties than other materials which are used in making biosensor and the researchers are interested in using CNT for next-generation of sensors. Because of these properties the researchers consider that CNT biosensor has the potential of revolutionizing the sensor area.

The advantages of CNT biosensors help it to perform better in many of the biomedical sensing applications. Therefore, CNT-based biosensors are highly suitable as implantable sensors.

In some studies, the dynamic parameters of biosensor such as response time and sensitivity with either carbon nanotube or without carbon nanotube were investigated. Decreasing of response time and increasing of sensitivity because of increasing electron transfer rate in the presence of the CNTs were reported [44-47]. Moreover, CNT's have excellent catalytic activity which decrease their oxidative potential to avoid fouling problems. The enzymes can be chemically immobilized to materials in the presence of CNT. For all these advantages of CNT biosensors are very convenient device to detect biological molecules.

CNT based Electrochemical Enzymatic Biosensors:

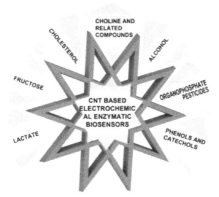

Fig. 13. The Types of Carbon Based Electrochemical Enzymatic Biosensors

Glucose biosensor: Nowadays, the glucose biosensors have an important role for diagnostic and control of diabetes. There are many ways of preparation of glucose biosensors which are made up of carbon nanotubes (CNT). For example, Rubianes and Rivas modified Carbon Nanotube Paste Electrode (CNTPE) with glucose oxidase (GOx). They obtained more sensitive glucose biosensor without redox mediators, metals etc. [26, 48].

The other atractive preparation of glucose biosensor is cross-linking of GOx with SWCNT and poly[(vinylpyridine) Os(bipyridyl)$_2$Cl$^{2+/3+}$] polymer film following two alternatives as shown in figure which is done by Schmidtke and co-workers. The first alternative is the SWCNT which was deposited on bare glassy carbon electrode (GCE) and then hydrogel containing the redox polymer and the enzyme for catalytic effect. Second alternative is SWCNT which were developed with enzyme solution after this process it was treat redox hydrogel and then modify with GCE [26, 48-49]. (Fig. 14)

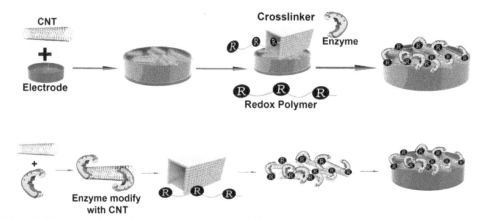

Fig. 14. The example of glucose biosensor based CNT. [35]

b) Fructose biosensor: Fructose is widely distributed monosaccharide and important sweetener because of its sweetening ability. The sweeting ability of glucose and sucrose is lower than that of fructose. The some scientists proposed amperometric biosensor, which was modified with CNTPE for sensing fructose, covered by polymer which is obtained from electropolymerization of dihydroxybenzaldehyde [26, 50].

c) Chlosterol A: The determination of cholesterol levels is of vital importance for some illnesses such as cardiovascular diseases. Chlosterol Biosensor consists of the modification of a screen printed electrodes (SPE) with chlosterol esterase, peroxidase, oxidase and MWCNT was used for determination of total cholesterol in blood with very good sensitivity. The Chlosterol biosensors modified with the carbon nanotubes promoted the electron transfer so as to improve the sensitivity of the sensor [26, 51].

The other biosensors, which are modified with CNTs, are Lactate biosensor, Phenols and catechols, Hydrogen peroxide, Alcohol biosensor, Choline and related compounds, Organophosphate pesticides.

CNT-based-DNA biosensors: The traditional method for sensing DNA and RNA molecules is too slow and requires special preparation. There are some critical points for preparation of DNA biosensor. Most important step is immobilization of DNA probe on the electrode. Media should have special conditions for pH and temperature for preparation of DNA biosensor. [26, 52]

Ye and Ju developed the use of SPE modified with MWCNT. The scientist developed the use of SPE modified with MWCNT. Their DNA biosensor detected the analyte fast and provided sensitive measurement [26, 53].

Fang and co-workers was used a glass carbon electrode (GCE) modified with MWCNT observed an enhanced sensitivity for electrochemical DNA biosensor based on carbon nanotubes. [26, 54]

DNA detection sensitivity of biosensor which is modified with CNT is much higher than conventional DNA sensors [26, 55].

2.4 Protein engineering for biosensors

What is protein engineering?

Protein engineering is the process of controlling the development of useful or valuable proteins. Proteins were used for specific biosensor design. Affinity between protein and analyte is the basic principle of this study area. Scientists, firstly determine the three dimensional crystal structure of the proteins and build a protein data bank. Three dimensional structures of the proteins are obtained with protein crystallization methods. When proteins are immobilized on the electrode surface, the active site of the proteins should be free in three dimensional structures. In some situations mutations can be applied to the active site of the proteins. Therefore, protein structures should be well known.

The interaction between protein and its ligand is determined with different types of transducers. If the presence of very low amounts of biomolecules is determined, various diseases and cancer types can be identified at early stages. Protein engineered biosensors can specifically identify chemical substrates with protein-based sensors. There are three main strategies employed in the engineering of more suitable biological components used in biosensors. These techniques do not exclusive to each other, also they can be applied together. Rational protein design, directed evolution and de novo protein design are the main methods. Each design strategy has limitations, advantages and disadvantages respect to each other to be used in a biosensor format. The three design techniques are used to modify aspects of stability, sensitivity, selectivity, surface tethering, and signal transduction within the biological environment [50].

Rational design of proteins

In rational protein design, the scientists use detailed knowledge of the structure and function of the protein to make desired changes, since site-directed mutagenesis techniques are well-developed. This has the advantage of being inexpensive and technically easy. However, detailed structural knowledge of a protein is often unavailable. When it is available, it can also be extremely difficult to estimate the results of various mutations. Computational protein design algorithm aims to identify amino acid structure sequences. While the conformational sequence structure in the space is large, a fast and accurate energy function is required that it can distinguish optimal sequences from similar suboptimal ones.

Directed evolution:

In directed evolution, mutagenesis method is applied on a protein, and a selection way is used to pick out variants that are quality. This method mimics natural evolution and generally produces superior results to rational design. An additional technique known as DNA shuffling mixes and matches pieces of successful variants in order to produce better results. This process mimics the recombination that occurs naturally. The most important advantage of directed evolution is that there is no need to know structure of a protein, and predict the resultant effect of a mutation. In fact, the results of directed evolution experiments are often surprising. Because the desired changes are often obtained by mutations that were not expected to have that effect. Disadvantage of the method is low throughput. This is not convenient for all proteins [51].

Future Biosensors Directions:

Miniaturization of developed biosensors will be important in the future. Because miniaturization is required for small electrodes, for example measurements in vivo. Another future approach is the combination of biological materials with a silicon chip because it seems to be the most comprehensive integration between biology and electronics [50]. Nanostructures will be important new components in recently developed electrochemical biosensors: Nanowires, carbon nanotubes, nanoparticles and nanorods are some of the familiar objects that are crucial elements of future bioelectronics devices and biosensors [52].

2.5 Wireless biosensors networks

An Aspect of Sensor Communication Networks

Each sensor or device communicating each other and a center with hierarchical protocols and/or functioning algorithms can be defined as a network. Network system which has either wired or wireless network system can access these sensor or device with a path. Even though wireless systems have become common with recent effective developments; some applications require wired network system. Beside this; topology means how network systems connect and operate. Each network system has its own topology. In other words; it is network architecture and is all efforts on hierarchical communication and functions between network members. Also it can be said that it realizes operation protocol (software).

For example; Ethernet is a network topology and TCP/IP is an access protocol. Topology also defines maximum access distance. While physical topology describes how the networks connected each other; logic topology describes how the network members transmit data. Following are some types of networks; commonly used I^2C and CAN BUS network topology; wireless networks and wireless sensor/biosensor network (WBNS). The system included point to point communication in earlier generation networks and the sensor included point to point communication, the sensor was communicating to a center. This communication was developed in 1980's. There were two main problems such as wave quality and cost. Wave quality was not enough and the cost was very high. After that in 1990's networks began to use micro controllers and some kind of sensor processor systems. Generally an analog signal come to this system then it is converted to digital signal and saved signals transmitted by RS 232, RS 422 or 485 protocols. Normally RS 232 works with binary code. The connections of the signal are made from datas to terminal equipment at the same time data circuit terminating equipment [53]. Logic signal is zero or one respectively for zero (+3) - (+15) V and for one, (-3) - (-15) V.

The smart sensor networks use bus system. Bus systems include bus connection system and bus system hierarchical protocol. Whole bits have two open ends. The data speed is 100Kbit/s at standard mode, 400 Kbit/s and 3,4 Mbit/s respectively fast and high speed wave mode. In order to compare to these bus systems, three bus systems are given in below. One of them is inter – integrated circuit I^2C bus, others are CAN BUS and Ethernet Bus protocols of network system topology [53, 54, 55].

I^2C bus network given in Fig.14 is suitable for sensing in short distances and process. Fiber optic or coaxial cable travels between network members in order and data is transmitted to all system at the same time thorough them. The fiber optic cables are using wired fiber optic sensor network. Since the fiber optic cables have multi wires, data is transmitted very fast. In wired sensor networks require fast transmission; electromagnetic waves are produced. In those networks; each member can send data to network because each member has unique MAC address. Addressed networks can decide whether the data travelling data is belonging to it or not with the help of the MAC number. Each member can manage it through the software on the network [56].

The devices in a network (actuator, sensor, or group of sensors, hd (eeprom) or a processor) can communicate each other. It is impossible to extend the device connection with this generation of sensor networks. In the case of using RS-232; extension problem is relaxed a bit. But usage of finite number of members must be mentioned. When data command send to the related device, the master one pass on receiver mode. Then related device become sender and sends the data to data line. [57] Clock is for reading the data. Bus systems consist hardware and a suitable protocol. In Fig.15 presents the processor sending data to devices (and nodes) with it is protocol.

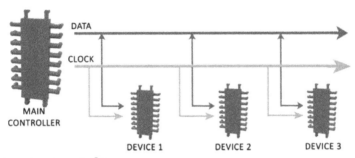

Fig. 15. The bus structure of I^2C

CAN bus is a differential system and works differential if high (logic 1) bigger than low (Logic 0). The differential value is 1. Opposite of this the value is naturally 0. This knowledge can be reached up to 1000 m. RS 422 and RS 485 is also differential. It is used in communication to reach points. The software supports the hierarchical protocol of the system. I^2C and RS 232 protocols are not differential and can be established a network only for short distance communication [58], [59]. Separately CAN differential system has good noise immunity and this system is secure with this side. Can bus system is given below in Figure 16.

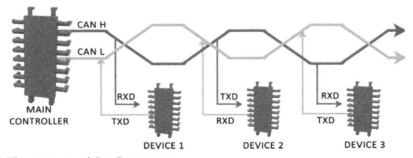

Fig. 16. The structure of Can Bus system.

In last decade of 20th century, wireless sensor networks became effective and nowadays it is going on process. This system has several active nodes. A node may include any sensitive device, processor, calibrator, power supply, software and cryptos. Each node of network has sensor or device and the communication of them is made with route algorithms that continuously developed. These networks are named as WBSN for biosensor application. Micro electro mechanic (MEMS) is also affecting the development of WBSN technologies [53]. In such cases to define new area for sensor network development, some hybrid systems and different hierarchical routine protocols are needed to be improved [58- 60].

Wireless Sensor Biosensor Networks

In many applications, getting and monitoring information in wired way from sensors is not possible. Therefore, the values of these sensors with wireless perception monitoring methods are needed. When wired sensor networks compared with wireless sensor networks, WSBN and WSN have good opportunities about security of these networks with using cryptos, limitations, robust systems e.t.c. [59], [61].

On the other hand, data transmission, communication costs and low power consumption are remarkable aspects of wireless sensor networks. Some of the several nodes includes functioned nodes and information systems are evaluating in the same system at the same time with the development of detection (sensing) technology and sensors in many applications in many areas. Especially data coming from multiple points and small functioned nodes increases the accuracy to be perceived. So, complex networks system has sustainable configuration.

Therefore wireless communication has been a hot topic nowadays. WSN (wireless sensor network), the name of this book, the WBSN (Wireless Biosensor Networks) is a wireless network system which continues its development since the 1990's. These network systems have widespread application areas such as military, pollution observation, natural disaster, healthcare etc. Medical imaging, medicine releasing, remote sensing, remote measuring, mine detection, wild life observations technologies have been developed which requires WBSN.

The network is in communication with each other consists of a large number of nodes as shown in Figure 17. This network has a wide coverage area and in tolerating errors but sometimes having limited computational and memory capabilities as depending on the topology of sensor networks and communication overlapping may prevent usage of this system.

A wireless network grouped the nodes in clustering and some file algorithms. Some clusters and trees include some sub clusters. İt ensures the better communication performance in getting and processing datas that comes from biosensor network. This clustering system works on a base. Each cluster has a head named gateway and includes maximum 5 grouped levels. Gateway gives performance to reach to whole sensor in own cluster. These clusters includes sub cluster in a tree structure. Some route algorithms restricts the number of sub groups. Each cluster has a main point named as gateway. Whole system that consist mainly of gateway is a WSN.

In order to operate and organize the system an interface runs. This interface depend on operation protocols that chosen. Clustering systems is built as some models that given in literature work reach system. The communication protocol are a software and communicates not only cluster via gateway but also to the sensor or device. The relaying and sensing is carried on sensor nodes [62],[63].

Wider-open systems can be found using these systems with classical internet, satellites and other networks. Sensors can communicate with each other each other using software. According to certain criterions but any nodes don't have any information in other nodes. This is the principle of any network. The wireless biosensor network provides access to the information easily anytime, anywhere [61].

A transceiver device and an actuator/device that gives or performs control commands are also available together with the sensors in network in these nodes for structuring of perception. It can be said that these nodes are physically in the same structures. The data stream and processes in this system is usually carried out by a process called. If these datas are analyzed by using different criteria and calculation algorithms, they will be transmitted to a central system from this base. All of these processes are handled by a network protocol and hardware [64]. The system can be built is a WBSN software protocol which has a high accuracy and reproducibility. Required software and communication protocols should be installed for the operation of the network system. The main currency protocols achieve very high speed incoming and outgoing data traffic. Protocols can be separated Data-centric protocols, Hierarchical Protocols, Locationbased Protocols. [65].

Particularly, at all of WSN Technologies, the routing protocols are developed in fast way. Some articles proposed these protocols. The routing protocols designed for WSN/WBSN can be classified based on path selection, as proactive, reactive, and hybrid and so on. These type of developing routines can be found in literature detailed and can be seen state of art [66].

In order to perform sensor network application can be used wireless adhoc networking. An ad-hoc is a network and it works as local area network (LAN). This system supports the devices, sensor connect as adhoc query. The knowledge signals is relayed from each node to other node. An adhoc network can organize this message traffics without any router. Adhoc networking is well known procedure not only sensors and devices but also development of data transmission and electronical application methods is carrying out the ad-hoc network routing protocols in WBSN systems.

But this technic cannot also effort some applications [56], [53]. Otherwise it is also partly old technic which includes much more nodes that may cause to pass other networking systems [67], [55] [68], [69].

So, entire network system receives the data, processes them, and also analyses and transmits them. These processes are calibration, linearization of nonlinear data, etc. Some of the nodes are equipped only with a functionality of continuing these processes, and the others can also provide energy to the system. In this way, an intelligent system can be achieved and operated at a great extent. There can be some nodes which are not operating while the other components of the system keep operating [69].

The system shown in Fig.17 defines the overall flow of WBSN. Fig 17 is made of transmitter and receiver. The transmitter consists of pulse generator, A/D converter, amplifier, PN spreading, modulator and radio transceiver [57], [64].

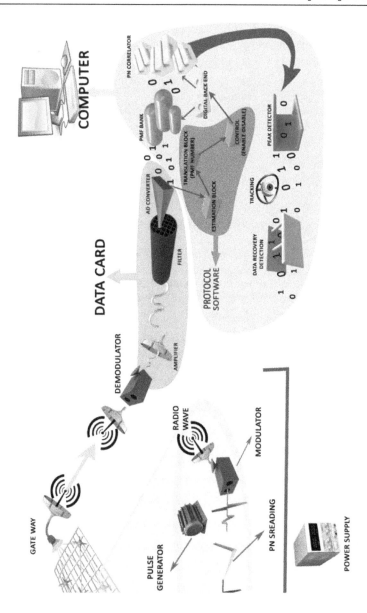

Fig. 17.

Data signal comes from a biosensor. Data is amplified and converted into digital by A/D converter. It is spread over a bandwidth by using PN Spreading and modulated by the modulator block. Modulated signal is then transmitted through the gateway by using Radio transceiver. Gateway in transmitter side takes all signals from all connected nodes and sends the signal to the gateway of the receiver. Receiver amplifies the signal, demodulates, filters and converts into analog by using A/D converter. Analog signal is placed to PMF bank. Digital backend operates both PMF bank and PN Correlator. The digital backend operator is fully controlled by Protocol software, which contains estimation block, translation block and Control block (enable/disable). As a result, data is processed by using data detection, tracking and data recovery detection. In this flow, computer obtains the signal blocks from sampling frequency block and signal tracking is done. Since signal location in PMF varies with time and this dynamic change is traced by Tracking. Therefore, stable data detection becomes possible.

Also, Computer takes the data from PMF Bank. Estimation block analyzes where each block is. Afterwards, data is clustered and configured in accordance with PMF Number. If the data is used, control block enables the signal. If not, it disables the signal. This control is managed by the software.

Security of Wbsn

Security is one of the most important issues for WBSN systems since the application areas of these systems are highly sensitive. For example, any intrusion to a system in military or automotive industry can be directly related to life safety. Security of the system is as much important as its sustainability. Security of WSN/WBSN systems can be improved using more powerful protocols and crypto. So, a variety of forms are going to be exhibited in the design and adaptation of networks. These requirements are going to provide forming of architectural designs.

For the security of data and network; using new operational algorithms for nodes without external influences and threats, confidentiality of data, providing long lasting network operations, gathering the data without delay, broadcast and multicast identification systems are the other fields which help to improve the security [70], [71].

The Radio Characteristics of Wbsn Communication Systems

Systems WBSN's data applications need communication rates from Kbps to Mbps. Average distance of sink/base station is between 10 and 30 m. The signal that comes from sensor in any node is transmit to the center of Adhoc system as WSN/WBSN. A node gets a decision and this knowledge does not depend on network main communication and structure. High speed of data transfer is required especially for image transmission in WSN/WBSNs.

Ethernet cards transfer data at 0.1 GHz (100 Mbit). Thus, a bit per 10 ns data pulse is achieved. Following can be explained about the available radio waves: AM (amplitude modulation) radio waves called as the carrier have 400 kHz frequency and transmit the data. Data is modulated as amplitude. Receiver filters the data with 400 kHz and converts the data from the amplitude of the wave which is known as demodulation. AM is not a good choice for modulation. Because noise is embedded into the carrier, and so the quality of the data obtained from demodulation is poor. Data can be transmitted to far distances, but the effect of noise is elevated.

Frequency modulation (FM) is another system that can carry data and transmit over a carrier wave with modulation in frequency and this frequency is constant but amplitude of the carrier changes. This is different to amplitude modulation method. FM baseband is between 88 and 108 MHz. Frequency of the carrier is varied into UHF (ultra high frequencies), VHF (very high frequencies) and UWB (ultra wide range) bands. Therefore, different frequency bands are occupied for various data migration.

GSM has the highest frequency band and data is carried as package-to-package in GHz level. In WSN, error-free data transmission is important. Depending on how far and how fast the data is transferred, wavelength must be selected. For example, a specific frequency range is assigned in satellite communication. Because, noises introduced in atmosphere levels must be separated. LNB (Low-noise block) can get very small signals. Reflected waves are focused on LNB and amplified to a sufficient level to process.

In addition to these, another network called "PHS" is commonly used in Japan mobile communication. PHS network provides good data communication in isolated locations such as underground and tunnels. Therefore, PHS is an option to use in WBSN under difficult environments. Available frequency bands for WBSN is from 6765 kHz up to 246 GHz [61].

3. Acknowledgement

Thanks to Enes ADANIR and Mehmet İŞCAN for nice pictures in our chapter and also thanks to Kadriye ATICI KIZILBEY, Nural PASTACI and Assoc. Prof. Afife Binnaz HAZAR YORUÇ for their support and help.

4. References

[1] Sassolas A, Blum L.J, Leca-Bouvier B.D (2011) Immobilization Strategies to Develop Enzymatic Biosensors Biotechnology Advances 30(3): 489-571.

[2] Nambiar S, Yeow J.T.W (2011) Conductive Polymer-Based Sensors for Biomedical Applications Biosensors and Bioelectronics 26: 1825–1832.

[3] Thevenot DR, Toth K, Durst R.A, Wilson G.S (1999) Electrochemical Biosensors: Recommended Definitions and Classification, Pure Appl. Chem. 7: 2333-2348.

[4] Sadana A. (2006) Binding and Dissociation Kinetics for Different Biosensor Applications Using Fractals, Pages 219-242.

[5] Liu G, Lin Y (2005) Electrochemical Sensor for Organophosphate Pesticides and Nerve Agents Using Zirconia Nanoparticles as Selective Sorbents, Anal. Chem. 77: 5894-5901.

[6] Dixon B.M., Lowry J. P., O'Neill R.D. (2002) Characterization in vitro and in vivo of the oxygen dependence of an enzyme/polymer biosensor for monitoring brain glucose Journal of Neuroscience Methods Volume 119, Issue 2 Pages 135 - 142

[7] Arora P, Sindhu A, Dilbaghi N, Chaudhury A (2011) Biosensors as Innovative ools for the detection of food borne pathogens", Biosensors and Bioelectronics 28 (2011) 1–12.

[8] Kuila T, Bose S, Khanra P, Mishra A.K, Kim N.H, Lee J.H (2011) Recent Advances in Graphene-Based Biosensors, Biosensors and Bioelectronics 26: 4637– 4648.

[9] Kissinger P, Heineman W.R (1996) Laboratory Techniques in Electroanalytical Chemistry, Second Edition, Revised and Expanded (2 ed.). CRC Press, ISBN 0824794451.

[10] Allen B.J, Faulkner L.R (2000) Electrochemical Methods: Fundamentals and Applications (2 ed.). Wiley. ISBN 0471043729.

[11] Zoski, Cynthia G. (2007) Handbook of Electrochemistry, Elsevier Science. ISBN 0444519580.

[12] Bard, Allen J.; Larry R. Faulkner (2000) Electrochemical Methods: Fundamentals and Applications (2 ed.). Wiley. ISBN 0471043729.

[13] R.S Nicholson, ShainI (1964) Theory of Stationary Electrode Polarography. Single Scan and Cyclic Methods Applied to Reversible, Irreversible, and Kinetic Systems Analytical Chemistry 36(4): 706–723.

[14] Heinze J (1984) Cyclic Voltammetry Electrochemical Spectroscopy New Analytical Methods Angewandte Chemie International Edition in English 23 11: 831–847.

[15] Eggins B.R (2002) Chemical Sensors and Biosensors, John Wiley&Sons Ltd, England.

[16] Gray L.G.S, Appleman B.R (2003) Eis: Electrochemical Impedance Spectroscopy, Journal of Protective Coatings & Linings 2: 66-74.

[16] Du D, Chen S, Cai J, Zhang A (2007) Immobilization of Acetylcholinesterase on Gold Nanoparticles Embedded in Sol–Gel Film for Amperometric Detection of Organophosphorous Insecticide, Biosens. Bioelectron. 23: 130–134.

[17] Biran I, Walt D.R (2004) Optrode- Based Fiber Optic Biosensors (Bio-Optrode). In: Ligler F.S, Taitt C.A.R, editors. Optical Biosensors Present and Future. Elsevier, Amsterdam: ISBN:0-444-50974-7. pp 5-16.

[18] Eggins B.R (2004) Chemical Sensors and Biosensors, West Sussex PO19 SSQ, England: ISBN 0 47 1 899 13 5 (cloth) 0 71 89914 3. pp. 51-52.

[19] Marazuela M.D, Moreno- Bondi M.C (2002) Fiber Optic Biosensors-an Overview, Anal. Bioanal. Chem. 372: 664-682.

[20] Canh T.M (1993) Biosensors Chapman&Hall an Masson Paris: ISBN 0 412 48190 1 pp. 126-129.

[21] Aboul-Enein H.Y, Stefan R.I, Van Staden J.F, Zhang X.R, Garcia- Campana A.M, Baeyens W.R.G (2000) Recent Developments and Applications of Chemiluminescence Sensors, Critical Rev. Anal. Chem. 30(4): 271-289.

[22] Gübitz G, Schmid M.G, Silviaeh H, Aboul-Enein H.Y (2001) Chemiluminescence Flow-injection Immunoassays, Critical Reviews in Analytical Chemistry. 31 2: 141-148.

[23] Wang P, Liu Q (2011) Biomedical Sensors and Measurement, Newyork: Springer Zhejiang University Press, ISBN 978-3-642-19524-2.

[24] Feng M, Han H, Zhang J, Tachikawa H (2008) Electrochemical Sensors Based on Carbon Nanotubes. In: Zhang X, Ju H, Wang J, editors. Electrochemical Sensors, Biosensors and Their Biomedical Applications. Elsevier Inc. pp. 462-463.

[25] Ajayan PM (1999) Nanotubes from Carbon, Chem. Inform. 30 39: 1787-1799.

[26] Rivas G.A, Rubianes M.D, Rodr'ıguez M.C, Ferreyra N.F, Luque G.L, Pedano M.L, Miscoria S.A (2007) Carbon Nanotubes for Electrochemical Biosensing, Concepci´on Parrado, Talanta. 74: 291–307.

[27] Iijima S (1991) Helical Microtubules of Graphitic Carbon, Nature 354, 56 p.

[28] Ajayan P.M, Zhou O.Z (2001) Applications of Carbon Nanotubes in Carbon Nanotubes. In: G. Dreselhaus, Ph. Avouris (Eds.), Springer, Heidelberg, pp. 391–425.

[29] Dresselhaus M.S, Lin Y.M, Rabin O, Jorio A, Souza Filho A.G, Pimenta M.A, Saito R, Ge G, Samsonidze G, Dresselhaus G (2003) Nanowires and Nanotubes, Mater. Sci. Eng. C 23: 129-140.

[30] Kingston C.T, Simard B (2003) Fabrication of Carbon Nanotubes, Anal. Lett. 36 (15): 3119 – 3145.

[31] Zhou O, Shimoda H, Gao B, Oh S, Fleming L, Yue G (2002) Materials Science of Carbon Nanotubes: Fabrication, Integration, and Properties of Macroscopic Structures of Carbon Nanotubes, Acc. Chem. Res. 35 (12): 1045-1053.

[32] Balasubramanian K, Burghard M (2005) Small 1, 180 p.

[33] Iura H.H, Ebbe sen T.W, Tanigaki K (1995) Adv. Mater. 7, 275 p.

[34] Britto P.J, Santhanam K.S.V, Alonso V, Rubio A, Ajayan P.M (1999) Carbon Nanotube Electrode for Oxidation of Dopamina, Adv. Mater. 11, 11 p.

[35] Joshi P.P, Merchant S.A, Wang Y, Schmidtke D.W (2005) Amperometric Biosensors Based on Redox Polymer-Carbon Nanotube-Enzyme Composites, Anal. Chem. 77 : 3183-3188.

[36] Merkoci A (2006) Carbon Nanotubes in Analytical Sciences, Microchim. Acta, 152: 157-174.

[37] Banks C.E, Crossley A, Salter C, Wilkins S.J, Compton R.G (2006) Carbon Nanotubes Contain Metal İmpurities Which Are Responsible for the `Electrocatalysis' Seen at Some Nanotube-Modified Electrodes, Angew. Chem. Int. Ed. 45, 2533 p.

[38] Gooding J.J (2005) Nanostructuring electrodes with carbon nanotubes: A review on electrochemistry and applications for sensing, Electrochim. Acta 50: 3049-3060.

[39] Katz E, Willner I (2004) Biomolecule-Functionalized Carbon Nanotubes: Applications in Nanobioelectronics, Chem. Phys. Chem. 5: 1084-1104.

[40] Merkoci A, Pumera M, Llopis X, P´erez B, del Valle M, Alegret S (2005) New Materials for Electrochemical Sensing VI Carbon Nanotubes, Trends Anal. Chem. 24: 826-838.

[41] Wang J (2005) Electroanalysis 17, 7 p.

[42] Valc´arcel M, Simonet B.M, C´ardenas S, Su´arez B (2005)Present and Future Applications of Carbon Nanotubes to Analytical Science, Anal. Bioanal. Chem. 382: 1783-1790.

[43] Jiang M, Lin Y (2006) Encyclopedia of Sensors in: Grimes G.A, editor. American Scientific Publisher 6, 2: 25–51. ISBN 1-58883-058.

[44] Lu L.M, Zhang X.B, Shen G.L, Yu R.Q (2012) Seed-Mediated Synthesis of Copper Nanoparticles on Carbon Nanotubes and Their Application in Nonenzymatic Glucose Biosensors, Analytica Chimica Acta 715: 99–104.

[45] Wang Y, Du J, Li Y, Shan D, Zhou X, Xue Z, Lu X (2012) A Amperometric Biosensor for Hydrogen Peroxide by Adsorption of Horseradish Peroxidase onto Single-Walled Carbon Nanotubes, Colloids and Surfaces B: Biointerfaces 90: 62– 67.

[46] Hoshino T, Sekiguchi S, Muguruma H (2012) Amperometric Biosensor Based on Multilayer Containing Carbon Nanotube, Plasma-Polymerized Film, Electron Transfer Mediator Phenothiazine, Andglucose Dehydrogenase, Bioelectrochemistry 84: 1–5.

[47] Narang J, Chauhan N, Jain P, Pundir C.S (2012) Silver Nanoparticles/Multiwalled Carbon Nanotube/Polyaniline Film For Amperometric Glutathione Biosensor, International Journal of Biological Macromolecules 3: 672-678.

[48] Lin Y, Lu F, Tu Y, Ren Z (2004) Glucose Biosensors Based on Carbon Nanotube Nanoelectrode Ensembles, Nano Letters 4(2):191-195.

[49] Li J, Wang Y.B, Qiu J.D, Sun D.C, Xia X.H (2005) Biocomposites of Covalently Linked Glucose Oxidase on Carbon Nanotubes for Glucose Biosensor, Anal. Bioanal. Chem. 383: 918–922.

[50] Antiochia R, Lavagnini I, Magno F(2004) Amperometric Mediated Carbon Nanotube Paste Biosensor for Fructose Determination, Anal. Lett. 37: 1657-1669.

[51] Li G, Liao J.M, Hu G.Q, Ma N.Z, Wu P.J (2005) Study of Carbon Nanotube Modified Biosensor Formonitoring Total Cholesterol in Blood, Biosen. Bioelectron. 20: 2140-2144.

[52] Rivas G.A, Pedano M.L (2006) Electrochemical DNA Biosensors, in: Craig Encyclopedia of Sensors, American Scientific Publishers, ISBN 1-58883-059-4 3: 45–91.

[53] Ye Y, Ju H (2005) Rapid Detection of ssDNA and RNA Using Multi-Walled Carbon Nanotubes Modified Screen-Printed Carbon Electrode, Biosens. Bioelectron. 21: 735–741.

[54] Cai H, Cao X, Jiang Y, He P, Fang Y (2003) Carbon Nanotube- Enhanced Electrochemical DNA Biosensor for DNA Hybridization Detection, Anal. Bioanal. Chem. 375: 287-293.

[55] Cai H, Xu Y, He P, Fang Y (2003) Indicator Free DNA Hybridization Detection by Impedance Measurement Based on the DNA-Doped Conducting Polymer Film Formed on the Carbon Nanotube Modified Electrode Electroanalysis 9, 15: 1864–1870.

[56] Bergveld P (1996) The Future of Biosensors. Sensors and Actuators A: Physical: ISSN 0924-4247. 56: 65-73.

[57] Lambrianou A, Demin S, Hall E.A (2008) Protein Engineering and Electrochemical Biosensors, Adv. Biochem. Eng. Biotechnol. 109: 65-96.

[58] Brett C.M. A, Brett A.M.O (1994) Electrochemistry Principles, Methods, And Applications, Oxford University Press, Great Britain.

[59] Wang P, Liu Q (2011) Biomedical Sensors and Measurement, Zheijang University - Verlag Springler Press, 4.7: 183 - 195.

[60] Sichitiu M.L (2004) Cross Layer Scheduling for Power Efficiency in Wireless Sensor Networks.

[61] Akyildiz I.F, Su W, Senkorasubramanioam Y, Cayirci E (2002) Wireless Sensor Networks a Survey, Computer Networks. 28(4): 383-422.

[62] Miao L, Djouani K, Kurien A, Noel G (2012) Network Coding and Competitive Approach for Gradient Based Routing in Wireless Sensor Networks., Adhoc Networks., http://dx.doi.org/10.1016/j.bbr.2011.03.031

[63] Naik R, Singh J, Le H.P (2010) Intelligent Communication Module for Wireless Biosensor Networks. In: Pier Andrea Serra, editor. Biosensors. InTech. pp. 225-240.

[64] Manjeshwar A, Agarwal l D.P (2001) Teen: a Routing Protocol for Enhanced Efficiency in Wireless Sensor Networks. Proc. 15th Int. Parallel and Distributed Processing Symp. pp. 2009–2015.

[65] Heinzelman W, Chandrakasan A, Balakrishnan H (2000) Energy-Efficient Communication Protocol for Wireless Microsensor Networks. Proceedings of the Hawaii Conference on System Sciences pp. 1-10.

[66] Schurgers C, Srivastava M.B (2001) Energy Efficient Routing in Wireless Sensor Networks. Proc. Communications for Network-Centric Operations: Creating the Information Force. IEEE Military Communications Conf. MILCOM 2001. 1: 357–361.

[67] Cheng H, Yang G, Hu S (2008) NHRPA: a Novel Algorithm for Hierarchical Routing Protocol Networks Wireless Sensor, The Journal of China Universities of Posts and Telecommunications 15(3): 75-81.

[68] Prasad N.R, Alam M (2006) Security Framework for Wireless Sensor Networks. Wireless Personal Communications, 37: 455–469.

[69] Xu Y, Heidemann J, Estrin D (2001) Geography-Informed Energy Conservation for Ad Hoc Routing, Proceedings of the 7th Annual International Conference on Mobile Computing and Networking, MobiCom '01., 70–84.

[67] Akkaya K, Younis M (2005) A Survey on Routing Protocols for Wireless Sensor Networks, Ad Hoc Networks. 3, 325–349.

[68] Misra S, Dias Thomasinous P (2010) A Simple, Least-Time and Energy-Efficient Routing protocol with One-Level Data Aggregation for Wireless Sensor Networks, Journal of Systems and Software 83(5): 852–860.

[69] Perkins C, Das S.R, Royer E.M (2000) Performance Comparison of Two On-demand Routing Protocols for Ad Hoc Networks. IEEE Infocom 2000, 3 – 12.

[70] Intanagonwiwat C, Govindan R, Estrin D, Heidemann J, Silva F (2003) Directed Diffusion for Wireless Sensor Networking, IEEE/ACM Trans. Network. 11(1), 2–16.

[71] Karl H, Willig A (2003) A Short Survey of Wireless Sensor Networks, Technical University Berlin, Telecommunication Networks Group.

6

Medical Technology Management and Patient Safety

Mana Sezdi
Istanbul University
Turkey

1. Introduction

Health organizations are businesses in technological intensity. All hospitals contain several technologies such as magnetic resonance (MRI) technology, laser technology, X-ray technology, RF technology, micro-camera systems, ..etc. The management of these systems that is called as "biomedical technology" is not easy. The basic principle of biomedical technology management is based on safety. Especially, because the affected ones by the bad management are patients, the importance of the biomedical technology management is better understood. At this point, "patient safety" comes to the foreground. Patient safety is the most critical international issue because of countless bad-practice events also related to biomedical technology management.

Patient safety that is one of the basic components of quality of care is defined briefly as the patient is not damaged and is not exposed to medical errors. The main target in the patient safety applications is the establishment of a system to protect the patient from potential damage and to eliminate the possibility of error.

Health organizations should create the safe, functional and supportive environment for patients, relatives and employees. To achieve this goal, the physical environment, medical and other equipment and human resources must be managed effectively. Systematic steps in this regard will result in the expansion of patient safety practices. The main priority of accreditation programs is to ensure the safety of patients and employees. More than 50% of Joint Commission International accreditation standards are related to patient safety.

For quality studies focused on patient safety in health establishments, the safe use of medical technology is necessary. When the medical devices' qualifications, locations and preventive maintenance applications are considered, it is seen that the safety of medical devices is the most important issue to improve patient safety.

Factors affecting the patient directly in a hospital environment are medical devices and device users. High-tech medical devices that are used for the purpose of both diagnosis and treatment are the most important determining factors about patient safety. The design of medical devices, the interactions with each other and insufficient training of users are negative impacts on patient safety.

If these problems are examined one by one, it would be appropriate to consider the systematic errors that are impressive on patient safety. In this study, all these effects mentioned below will be handled one by one.

- Interactions between medical devices
- Sterilization
- Re-use of single use devices
- Medical device accidents and user errors
- The classification of clean room, particle measurement
- Radiation safety
- Electrical safety and
- Performance measurements of medical devices.

There are a number of recent studies about patient safety. Some of them are the surveys of medical device accidents and user errors (Carol, 2003; Hijazi, 2011; Brennan, 1991; Sawyer, 1997; Sezdi, 2009a). Some of them are focused on only electrical safety (Barbosa et al., 2010; Osman et al., 1996; Chakrabartty et al., 2010; Bakes, 2007; Sezdi, 2009b), whereas the others examined both the sterilization and reusage of single use devices related to the patient safety (Yoleri, 2011; Rice et al., 2009; Quirk, 2002; Northrup, 2000; Day, 2004; Koh, 2005; Buchdid Amarante, 2008; Hailey et al., 2008). There are also studies that explain the classification of clean rooms by measuring particles (Sezdi, 2009c). It is essential to collect all issues in order to ensure the safety management of medical devices.

2. Interactions between medical devices

The medical devices affect each other because of several interactions. Particularly, in the environments where many medical devices are connected to the patient such as operating rooms, intensive care services,…etc., there may be interaction between the devices. Especially, because of interference from radiofrequency energy that is called radiofrequency interference (RFI), there are several failures that cause to serious injuries and death. The reasons of the radiofrequency interference are the increasing numbers of electronically controlled medical devices with inadequate electronic protection against RFI and the increasing numbers of radiofrequency sources in the environment.

Mobile phones that are today's indispensable communication tools, enhance the potential for radiofrequency interference and have negative impacts on medical devices especially on pacemakers, apnea monitors and ventilators. The impacts of mobile phones with different frequency and UHF radio frequency receiver / transmitter on ventilators were examined in many statistical studies and a highly interaction was observed (Gilligan et al., 2000; PMDA, 2006; Lawrentschuk et al., 2004, Tan et al., 2001; Bassen, 1998; FDA, 1997; Carranza et al., 2011; Pressly, 2000; Hans et al., 2008).

For example, when a physiological monitor is used in conjunction with an electrosurgical unit, the ECG or arterial blood pressure waveform disturbs. Burns under the ECG electrodes may occur. During cardiac catheterization, ventricular fibrillation may occur.

If an anesthesia machine is used in magnetic resonance room, it creates the image distorts because its metal parts cause magnetic effect.

For the elimination of the negativity, the electromagnetic compatibility (EMC) standards related to medical devices and their placement in hospitals should be created. During new devices are purchased, electromagnetic compatibility should be considered.

3. Sterilization

Sterilization is the other important issue for patient safety because all materials in contact with mucous membranes must be sterile. Sterilization is a process to destroy all microorganisms found in or on a substance. In 1995, sterilization is defined by Association for the Advancement of Medical Instrumentation (AAMI) as a process intended to remove or destroy all viable forms of microbial life, including bacterial spores, to achive an acceptable sterility assurance level (AAMI, 1995). Sterility Assurance Level (SAL) can be defined as the probability to remain only one live sport when sterilization process is repeated a million times.

Although there are a lot of sterilization method, the most widely used sterilization method in a hospital environment is heat sterilization. The action mechanism of heat sterilization is to destroy proteins in the cell directly. In heat sterilization, the effective factors are temperature, time of the heat effect, the degree of moisture, the water content of microorganisms, pH and osmotic pressure.

Heat sterilization is classified as dry heat sterilization and steam sterilization. For effective dry heat sterilization, 175 °C process of 1 hour or 140 °C process of 3 hours is sufficient. Because of lack of humidity in environment, sterilization takes longer. In this manner, glass and metal instruments, oil and dust are sterilized.

Steam sterilization is performed at 121 °C under 1.5 atm pressure for 15 minutes. Usually, the materials that do not deteriorate under heat and pressure, are made sterile. For steam sterilization, the tool called autoclave is used. The basic principle is that every point of the material to be sterilized comes into contact with saturated water vapor in adequate time. For this reason, there are some principles to be followed during the packages are created and are placed autoclave. These include:

- The material to be sterilized must be cleaned by applying pre-cleaning process and it must be free from visible dirt.
- Packet size must comply with the size of the autoclave.
- Packaging material must be able to allow access and exit of steam.
- Non-woven packaging materials (plastic polymers, cellulose fibers and a specially produced paper) should be preferred because of their small pores.
- Labels reporting the contents of the package must be affixed on the packages.
- Date of sterilization, shelf life and department must be written on the label.
- Labeling process should not hurt the package.
- Packages should be placed loosely in the autoclave to contact the steam at each points.
- The materials that are not required packaging should be placed in the appropriate baskets or containers.
- The tubes filled with fluid material to be sterile must not be loaded completely.
- For the sterilization of materials in screw cap bottles, the caps should be loosened.

Material in the packages after the sterilization process must remain sterile until used. Sterile exposure time depends on the quality of packaging material, storage conditions, transfer conditions and the amount of materials. To maintain sterility after sterilization of the material, the considerations are as follows:

- Sterile equipment in case of moving and storing a long period of time, should be covered with a clean dust cloth.

- Sterile storage area must be close to sterilization area and entry to this room must be limited.
- Storage area must have a ventilation system for the appropriate temperature and humidity conditions. Temperature should be 18-22 ºC and humidity should be 35-50%.
- Packages should be kept in closed cabinets.
- If there is a damage on the package after sterilization, the material in the package should not be considered sterile.

In health care facilities, the devices that require more attention are sterilization devices. Because the materials and products in almost every department of the hospital have to be sterilized, sterilization devices and process must be controlled in intensity (Rutala et al., 2004; Dubois, 2002; McDermott, 2010; Kelkar et al., 2004). There are international standards in order to conduct inspection of the sterilization process. The standard of ANSI / AAMI ST79:2006 "Comprehensive guide to steam sterilization sterility assurance in health care facilities" recommends developing and implementing procedures for sterilization (ANSI/AAMI ST79, 2006). Sterility control is carried out in 3 main groups as physical control, chemical control and biological control.

Physical control is the control whether the autoclave unit works at proper temperature, pressure and time, or not. It is necessary to control the indicators of the device by the user before each use, and to perform maintenance service and validations by the biomedical staff periodically. Physical control gives us only information about the operation of the device, no information about the success of sterilization. Special designed temperature and pressure dataloggers are used for physical control of autoclaves (Figure 1). These are resistant to high temperature and pressure.

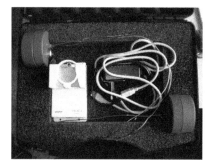 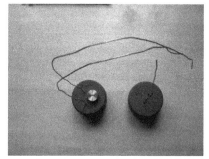

Fig. 1. Temperature and pressure dataloggers for physical control of autoclave (Yoleri, 2011)

As an example, the results of the physical control of an autoclave for the sterilization conditions of 121 ºC, 1,5 atm and 15 minutes, can be seen in Figure 2.

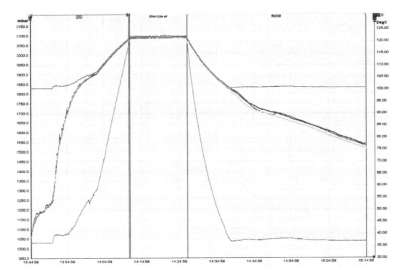

Fig. 2. The measurement results for the sterilization conditions of 121 °C, 1,5 atm and 15 minutes (Yoleri, 2011). Sterilization process generates where the fixed curve during maximum pressure and max temperature.

Chemical indicators are used for chemical control. Chemical indicators are indicators which give information about the sterilization process by changing their color in response to chemical or physical conditions or changing from solid phase to liquid phase. A positive result on a chemical process indicator does not mean that the material is sterile, it shows that the required parameters of the sterilization process are fully implemented.

The mostly used chemical indicators are Bowie-Dick test apparatus. Bowie-Dick test apparatus consist of a series of vapor permeation layer barriers. A complete color change indicates the sufficient steam penetration (Figure 3(a), 3(c)). If the color change is non-uniform or there is no any color change, it means that autoclave has an air suspension failure (Figure 3(a), 3(b)) and it must be controlled.

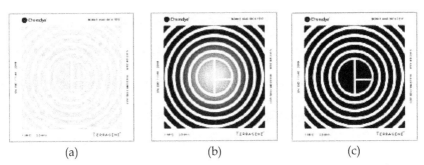

| (a) | (b) | (c) |

Fig. 3. The color changes on Bowie-Dick test apparatus (a) apparatus before application (b) non-uniform color change indicates insufficient steam penetration (c) uniform color change from yellow to black indicates sufficient steam penetration
(http://www.scu.com.tr/BOWIE&DICK%20TEST%20PAKET%C4%B0.pdf)

Biological indicators are used for biological control. In biological indicators, there are bacteria spores, known as the most resistant to sterilization. In the packet, biological indicators are placed in the points that are thought as the most difficult points for sterilization process. After sterilization, whether biological death has been ensured or not, is controlled by applying reproduce test.

Association for the Advancement of Medical Instrumentation (AAMI) suggests that the sterilized material to be implanted into the body should not be used until the reception of biological indicator test results.

When properly used, sterilization can ensure the safe use of invasive and noninvasive medical devices. The method of sterilization depends on the use of the medical device. To perform sterilization techniques successfully, current sterilization guidelines must be followed.

4. Re-use of single use devices

The other important problem in patient safety is the uncontrolled reuse of single use devices (SUDs). Although the single use devices are the devices that are used for one patient-one operation, it is possible to use them for second or more patient without any risk by applying the reprocessing technique. A "reprocessed device" is explained briefly as an original device that has previously been used on a patient and has been subjected to additional processing and manufacturing for the purpose of an additional single use on a patient (Rice et al., 2009). Reprocessing of medical devices may occur in a hospital or be contracted to a third-party reprocessing facility. About 25% of hospitals in the USA use reprocessed single-use devices in according to a survey by the Food and Drug Administration (FDA). 84% of the hospitals use third-party reprocessors to sterilize the devices, while 16% of them reprocess the devices in hospital (Quirk, 2002).

The approval of the reuse of single use devices is taken from FDA who is the authorized foundation. The reason of this application is that the manufacturers determine generally their devices as single use although the device can be used again after reprocessing. Manufacturers qualify their devices as single use devices because they want more production and more gain, and they want to launch their devices immediately without FDA approval because FDA approval time for reusable devices takes more time and manufacturers do not want to wait for a long time. Also, the manufacturer is not responsible to prove that a device can not be reprocessed. Because of this, although manufacturers say that the quarantee of working correctly of the reprocessed devices can not be given, FDA say that reprocessing is safe if FDA's requirements are performed (Northrup, 2000).

There are many studies about reusing of single-use devices. Some studies show that the reuse of SUDs is potentially safe and effective with reprocessing protocols and standards, while others do not recommend reprocessing and reuse because of the faulty devices (Day, 2004; Koh, 2005; Buchdid Amarante, 2008).

The FDA claims that the reprocessed device must be substantially equivalent to newly manufactured devices (Emergency Care Research Institute, 2006). It recommends that a medical device can not be reprocessed succesfully unless it can be cleaned, sterilized and function tested. Cleaning means removal of visible contaminants including body waste, body fluids, dirt, dust, etc. Sterilizing means meeting of domestic and international sterilization standards. Function tests verify that a device will perform as intended (Selvey, 2001).

Food and Drug Administration (FDA) has developed a list of known reprocessed SUDs (FDA, 2000). The list includes surgical saw blades, surgical drills, laparoscopy scissors, orthodontic braces, electrophysiology catheters, electrosurgical electrodes, respiratory therapy and anesthesia breathing circuits, endotracheal tubes, balloon angioplasty (PTCA) catheters and biopsy forceps. But, there is still uncertainty regarding the safety and effectiveness of the reuse of single use devices (Hailey et al., 2008).

Single use medical devices are classified by FDA according to the level and type of control needed to ensure that the devices are safe. Class I devices require the fewer controls, while Class II devices require "special controls". Class III devices are the most important class and are not adequate to reprocess.

Reprocessable Class I devices are devices that make contact and not penetrate intact skin. These are;

- General use surgical scissors,
- Non-electric biopsy forceps,
- Orthopedic chisels, knives and saw blades,
- Surgical curettes and gouges,
- Rasps.

Class II devices are devices that contact intact mucous membranes and not penetrate normally sterile areas of the body. These are;

- Laparoscopic scissors, clamps, dissecters and graspers,
- Compression sleeves,
- Recording and diagnostic EP catheters,
- Drills and burrs,
- Flexible snares.

Non-reprocessable Class III devices are devices that contact normally sterile tissue or body spaces during use. These are;

- Percutaneous and conduction tissue ablation electrodes,
- Transluminal coronary angioplasty catheters,
- Implanted infusion pumps.

During reusage of single use devices, two important risks should be considered. The first and the most important risk is the infection risk. The other risk is that the single use devices which are reused repeatedly, can not accomplish their function (Avitall et al., 1993, Rizzo et al., 2000, Zimerman et al., 2003). For example, in the reuse of catheters, some function risks may form. In the lumens, a contamination may occur, the catheters may slip and the distance between the electrodes may change. The worst of them, the electrodes may destroy the vessels by escaping from the catheter. Figure 4 shows the internal lumen of the reusable forceps having deformations after reprocessing.

To minimize the risks that affect patient safety directly, it must be required to obtain the information about the reuse of single use devices, to realize the applications standartly and to control the hospitals.

Fig. 4. Fig. 4. Photographs of the deformation in the lumen of the reusable forceps after reprocessing (Rizzo et al., 2000)

For patient safety, FDA proposed a strategy on reuse of single use devices. In the document that was published by FDA, the following steps are proposed to consider by hospital management (Henney, 2000).

- Regulation of third party reprocessors that reprocess SUDs in the same manner as the manufacturers of original equipment,
- Development of a device categorization system based on the risk level of SUDs,
- Writing comments on a draft list of frequently reprocessed SUDs,
- Providing information from the original equipment manufacturers on SUD's risk labels,
- Validation of procedure.

The testing procedures lead staff to develope specific protocols for cleaning, function testing and sterilizing of each reprocessed device. The testing procedures are known as validation.

Validation of reprocessing is classified as process validation and design validation.

- Process Validation controls the process consistently that produces a product meeting its predetermined specifications.
- Design Validation controls device specifications that conform with user needs and intended use.

A validation of reprocessing is necessary with respect to following safety issues (Popp et al., 2010):

- Physical safety (alteration of device's dimensions, weakening of components, poor performance, etc….)
- Chemical safety (absorption of cleaning agents, disinfectants, sterilization agents, toxic reactions, etc….)
- Biological safety (inadequate cleaning or disinfection of all surfaces, etc…..)

By doing validation studies, it can be proved that sterility will be achieved when the temperature and humidity parameters are used accurately. Because of this, validation studies must be established routine for each sterilization process.

5. Medical device accidents and user errors

Medical device accidents and user errors are problems that must not be ignored and need to focus on, because accidents occurred in the health sector cause patient death, or at least cause an injury and disability. According to the Food and Drug Administration (FDA), in the United States about 1,3 million people are injured each year by accidents caused from medical devices (Carol, 2003; Rados, 2004). The FDA receives more than 95.000 medical device accident reports annually (Hijazi, 2011). In 2002 alone, FDA declared the medical device accidents of more than 111.000 causing serious injuries and deaths (Carol, 2003).

In a study performed by Harward Medical Practice Study, it was explained that 70% of 30.000 medical device injuries are caused by medical device accidents while 27,6% of them are caused by pure negligence (Brennan, 1991).

Because the used devices are high-tech, accidents occur largely as a result of user error.

Many statistical studies show that user errors are not less. The user errors can be classified in related to hardware design, software design, components and alarms of devices.

The errors related to hardware include control/display arrangement errors. Especially, in the infusion pump display panel, the flow rate readout may be blocked from view. For example, the top of the 7 is blocked from view and it can be read as 1 (Sawyer, 1997).

Software-related design errors result generally from unfamiliar language, symbols, codes and functions that are hidden from the user. For example, in some cardiac output monitor alarm may be disabled without the operator's knowledge when the control buttons are pushed in a specific sequence (Sawyer, 1997).

The most common errors reported to FDA are improper installations of device accessories. Some commonly reported errors are tubing connected to the wrong port, loose connections and accidental disconnections. According to Medical Device Recall Reports, several injuries and deaths occurred because of the disconnections of the breathing tubes in the ventilators due to poor tube and connector design.

Alarm problems are false alarms, delayed alarms, too sensitive or insensitive alarms, inappropriate silencing and accidental disabling. There are many scenarios that cause patient death. Low alarm intensity, high ambient noise, low battery conditions, inappropriate alarm settings and other factors combine to create potentially dangerous situations.

The accidents in the anesthesia machine come the first in the medical accident list. The reason of the importance of the anesthesia machine accident is that the death of healthy patient that is operated for a basic problem is not accepted.

In operations, the electrosurgical unit may cause accidents because of the uncontrolled usage of patient electrode (damaged patient electrode or decreased patient contact). The other risks that might be caused from the electrosurgical unit, are the explosion of the anesthetic agent in the operation room and the possibility of the touching of the active electrode to the healthy tissues.

Fires that occur during operations are discussed as medical device accident. Although there are a lot of resources of the operating room fires, oxygen provider, area which collect ambient gas and the igniter are 3 points that should be considered. During the operation, each of 3 parameters is in the responsibility of different people in the operation team.

Oxygen provider system is controlled by anesthesiologist who monitors the patient during the operation. If too much oxygen, nitrous oxide, or other flammable gases accumulate under the drapes fabric used in the operation, a tiny spark may cause an explosion. To prevent the accumulation of oxygen or to control the amount of preparing solutions are the responsibility of nurse in the surgery, while the use of electrocautery or laser devices that could lead to the spark are under the control of the doctor. Because the 3 factors that could cause a fire during the operation are under the control of 3 different people, very good communication should be established in order to prevent fire. Lack of communication can cause a very big explosion or fire (AMN Healthcare, 2008).

There are many accidents that are caused from which MRI's magnetic field strongly pull all sorts of metal goods (scissors, hanger of saline physiology) near field (Figure 5). Although in some MR imaging centers, it is aimed to prevent such accidents by placing the metal detector at the entrance points, sometimes the cautions are skipped. In the patients, very rarely 2 or 3 degrees of skin burns resulting from contact with the magnetic field windings or cables may be observed (Sezdi, 2009a).

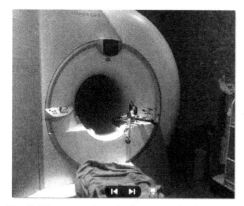 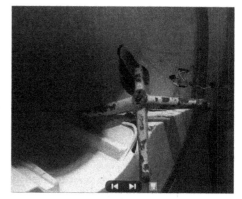

Fig. 5. Photographs of a hanger of saline physiology stucked to the MRI (Sezdi, 2009a)

"Risk Management" protocol within the scope of quality health care obliges to keep an absolute record of medical device accidents (Nobel, 1996). In order to prevent the repetition of these accidents, personnel training programs and security system are a contribution to the work of "patient safety", it can not be separated.

6. The classification of clean rooms, particle measurements

In hospitals, the operation rooms and intensive care rooms are known as clean rooms because it must have the high quality air without the dust and the particles. The clean rooms are closed places whose temperature, humidity, pressure and particles must be controlled.

In the clean room, the dust that is sourced from both the personnel and the patient and the machines, causes the risk of infection. Particularly, in the intensive care rooms for babies, the risk of infection is higher. By using the clean-air system, it is seen that the percent of infection risk decreases. But, this clean air system must be controlled by using the method of "the clean room classification" (Sezdi, 2009c).

Clean rooms in hospitals are divided into 2 groups in terms of particle concentrations. Class I rooms are areas that require high degree of hygienic conditions. These include:

- Operating rooms
- Sterilization rooms
- Sterile equipment storage rooms
- Intensive care patient rooms
- Newborn baby rooms
- Surgical department
- Patient rooms which is danger of infection

Class II rooms are areas that require normal hygienic conditions. These include:

- Patient rooms
- Emergency units
- Laboratories
- Caesarean section rooms
- Radiology, nuclear medicine treatment rooms
- Morgue and autopsy rooms

Operating rooms are divided into groups in itself as 1A and 1B. 1A-class operating rooms are operating rooms which require very high hygiene environment. In this type of operating rooms, air flow is done with laminar flow units which produce low turbulent air flow. The interventions in the Class1A-type operating room are as follows:

- Heart and vascular surgery
- Brain surgery
- Transplantation
- Bone marrow transplantation
- Orthopedic or interventions after accident

Class 1B operating rooms are used for interventions that do not require low-turbulent flow. The operations in the 1B-type operating rooms are as follows:

- Arthroscopy
- Thoracoscopy
- Laparoscopy
- Bronchoscopy
- Endoscopy
- Cardiac catheter examination

The classification of the operation room and the intensive care room is determined by the international standard of particle measurements. The related standard is ISO 14644-1:1999(E) "Cleanrooms and associated controlled environments Part 1: Classification of air cleanliness".

The main criteria for the classification of clean rooms is the particle dimension (0,1μm, 0,2μm, 0,3μm, 0,5μm, 1μm and 5μm) and the particle concentration. In according to this standard, the mean particle concentration from each point must be equal to the limit particle concentration or lower.

The maximum permitted concentration of particles for each considered particle size is determined from the following equation (ISO 14644-1, 1999):

$$C_n = 10^N \times (0{,}1 / D)^{2{,}08} \qquad (1)$$

where

C_n is the maximum permitted concentration (particles/m³ of air)
N   is the ISO classification number, which shall not exceed a value of 9.
D   is the considered particle size, in micrometers.
0,1 is a constant.

Table 1 presents selected particulate cleanliness classes and the corresponding particle concentrations.

ISO Classification Number (N)	Maximum concentration limits (particles/m³ of air) for particles equal to and larger than the considered sizes shown below.					
	0,1 µm	0,2 µm	0,3 µm	0,5 µm	1 µm	5 µm
ISO Class 1	10	2				
ISO Class 2	100	24	10	4		
ISO Class 3	1 000	237	102	35	8	
ISO Class 4	10 000	2 370	1 020	352	83	
ISO Class 5	100 000	23 700	10 200	3 520	832	29
ISO Class 6	1 000 000	237 000	102 000	35 200	8 320	293
ISO Class 7				352 000	83 200	2 930
ISO Class 8				3 520 000	832 000	29 300
ISO Class 9				35 200 000	8 320 000	293 000

Table 1. ISO classification numbers related to particle concentration (ISO 14644-1, 1999)

In clean rooms, making the control of the patient lying area such as operating table in operating room, patient beds in intensive care services, rather than controlling of HEPA filters is important. At that points all particles in the air must be sucked back to the ground and subdued.

For the particle measurements, a particle measurement test device is used. Measurements are taken from different points whose number is calculated from the square root of the area (m²) of the clean room. For example, in a clean room of 25 m², the measurements should be taken from 5 different points. The measurement results are analyzed by comparing the limit values in ISO 14644-1 standard.

Accordingly, although the clean room with laminar flow system should provide the conditions of ISO Class 3, the clean room with HEPA filter should provide them of ISO Class 5.

After validation, a full-fledged evaluation report including the main scheme and the sampling plan of clean rooms, measurement results and calibration dates of measurement devices, is prepared.

7. Radiation safety

During radiation, if the prevention is insufficient, both the patient and the user are exposed with high value dose and they are affected from high dose badly. Radiation doses cause severe destruction on the skin and harm to human health depending on the level of dose.

The application of Radiation Safety Program is important in accordance with safety of both patient and user. The objective of Radiation Safety is to provide standardization to prevent radiation's ill effect. The International Commission on Radiation Protection (ICRP), the International Atomic Energy Agency (IAEA) and other various independent institutions have been making publications in relation to ionizing radiation protection for more than fifty years. Report 60 of the ICRP and the Basic Safety Standards that was published in the IAEA report have three basic principles related to the radiation protection (ICRP, 1991; IAEA, 1996).

For medical radiation application, the prevention method from radiation can be summarized as the following three headings:

- Justification: do not permit the non-benefit radiation. By considering the harmful results of radiation, the radiation with clear advantage is only accepted.
- Limitation: limitation of annual radiation dose. The annual radiation dose must not be exceed the limit value.
- Optimization: to expose possible minimum dose. Except for radiotheraphy radiation, the minimum dose is exposed to the person by considering the economic and social factors.

The main two important issues of radiation management are, in general, the implementation of a quality control program for usage of radiation and for monitoring the quantity of doses received by individual patients, and continuous training of the device users.

For manageability of the radiation protection process, in each institution of radiation, a radiation safety program and certain precautions for patients who are treated with radioactive materials in unusual cases, must be followed.

Under this program, radiation safety training should take place in a serious way. A radiation safety manual should be prepared and if needed, should be presented as a resource accessible.

A committee related to radiation safety should be established, additionally effective and efficient operation of this committee should be provided. Principles of radiation protection should be ascertained and a procedure for application should be established. It is essential to control whether safety rules are applied or not.

Working with radiation, precautions should be applied carefully. In the walls of the rooms where all the devices that work with X-ray tube, as a minimum 1.5 mm thick lead shielding material should be used. Lead apron must be used in scopy room and must be used during shooting. Lead aprons are made from 0,25 to 0,5 mm lead equivalent material. Lead aprons must be stored by hanging for not broken, never fold it. When working close to the X-ray beam, equivalent lead thickness of protective gloves should be 0,5 mm. The doors must be kept closed during the shooting.

For the National Radiation Safety Program, each foundation that is related to the radiation must be controlled.

8. Electrical safety

The electrical safety is essential in patient safety because of electrical shock. In cases of electrical shock, the important thing is the electrical current flowing from human's body. The current not voltage is often the source of injury or death.

When an electrical current flows through the human body the effect of current is influenced by two main factors. Firstly the amount of current and secondly the length of time that the current flows.

The effect of current on the human body can be given as below (Webster, 1992):

- 0,5 - 10 mA threshold of perception
- 6,0 – 50,0 mA let-go current
- 75,0 – 400,0 mA ventricular fibrillation
- 1–10 A myocardial contraction, burns and physical injury

When the electrical current exceeds a certain limit value, the electrical shock that are called as "macroshock" occurs. In macroshock, the effect of current starts with a slight feeling on the skin. By increasing the value of electrical current, muscle cramps and spasms, difficulty breathing, ventricular fibrillation, burns and death occur.

In the invasive techniques that reduce or eliminate the resistance of skin, the patient is unprotected against electrical shock. ECG electrodes reduce the skin resistance because of the gel between them and increase the risk of electrical shock. However, intravenous catheters serve as a good conductor mounted directly to the heart when the contrast agent is used. As a result of leakage currents, electrical shock is seen in patients with catheter and called as "microshock" (Barbosa et al., 2010; Osman et al., 1996; Sezdi, 2009b).

To prevent patients from electrical shock, the electrical safety measurements of the medical devices should be performed by considering their different electrical specifications that are expressed by standard symbols (Figure 6). Medical devices are classified as Class I and Class II according to their electrical specifications. Class I devices are provided with basic insulation. Class I equipment is fitted with a three core mains cable containing a protective earth wire. Exposed metal parts on Class I equipment are connected to this earth wire (Chakrabartty et al., 2010). Class I is the most common type.

Class II equipment is enclosed within a double insulated case and does not require earthing conductors. Class II equipment is usually fitted with a 2-pin mains plug. Class II or double insulated equipment can be identified by the Class II symbol on the cabinet (Chakrabartty et al., 2010).

Because some medical devices have applied part which is designed to come into physical contact with the patient, electrical specifications of applied parts must be known for safety testing (Backes, 2007).

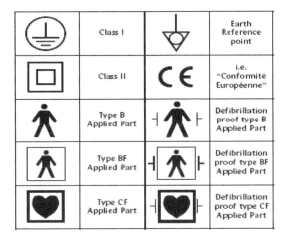

⏚	Class I	⏚	Earth Reference point
▢	Class II	C E	i.e. "Conformité Européenne"
🧍	Type B Applied Part	⊣🧍⊢	Defibrillation proof type B Applied Part
🧍	Type BF Applied Part	⊣🧍⊢	Defibrillation proof type BF Applied Part
❤	Type CF Applied Part	⊣❤⊢	Defibrillation proof type CF Applied Part

Fig. 6. The most commonly used symbols for electrical specifications of devices (Backes, 2007).

F-Type Applied Part is electrically isolated from Earth. F-type applied parts are usually used as either type BF or type CF Applied Parts.

Type B Applied Part is usually Earth referenced. Type B applications can not be used for direct cardiac application.

Type BF Applied Part has a higher degree of protection against electrical shock. But like type B, they can not be used for direct cardiac application.

Type CF Applied Part has the highest degree of protection against electrical shock. They are suitable for direct cardiac application.

In order to control the electrical safety of medical devices, there are some standards that were produced by the International Electrotechnical Committee (IEC). These standards are IEC 60601 and IEC 62353. Although these standards have some differences about the measurement technique of earthbond and leakage current, both of them are used for electrical safety analyses of medical devices (IEC 60601-1; IEC 62353).

Earthbond measurements are performed to check the low resistance connection between the earth and any metal parts. The resistance is measured between two probes. The first probe is connected to the earth point and the other is connected to the metal parts of the medical device.

Leakage measurements are performed from 4 different sources (Backes, 2007). These are; Earth Leakage, Enclosure Leakage, Patient Leakage and Patient Auxiliary Leakage. In according to the IEC 60601-1, the limits of leakage currents for different type applied parts can be seen in Table 2.

- Earth Leakage: It is the current flowing down the Earth conductor of the mains inlet lead (Figure 7).

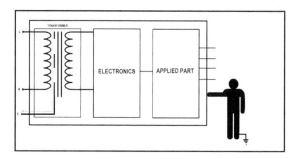

Fig. 7. Schematic diagram of earth leakage

- Enclosure Leakage: It is the current flowing to Earth through a person by touching the medical equipment (Figure 8).

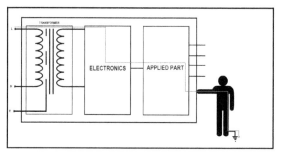

Fig. 8. Schematic diagram of enclosure leakage

- Applied Part or Patient Leakage: It is the current flowing through a person to Earth from the Applied Part or the current flowing from a person to Earth via the Applied Part (Figure 9).

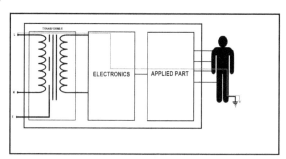

Fig. 9. Schematic diagram of patient leakage

- Patient Auxiliary Current: It is the current flowing between parts of the applied part through the patient (Figure 10).

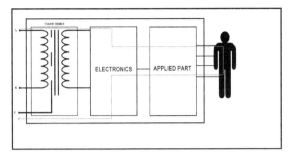

Fig. 10. Schematic diagram of patient auxiliary leakage

All leakage measurements are done by introducing conditions that could occur under normal use and under a single fault condition (SFC). The leakage tests are performed as follows:

1. Normal Supply Voltage
2. Normal Supply Voltage Open Neutral
3. Normal Supply Voltage Open Earth
4. Reversed Supply Voltage
5. Reversed Supply Voltage Open Neutral
6. Reversed Supply Voltage Open Earth

	Type B Applied Parts		Type BF Applied Parts		Type CF Applied Parts	
Leakage Current Type	NC	SFC	NC	SFC	NC	SFC
Earth Leakage	0,5mA	1mA	0,5mA	1mA	0,5mA	1mA
Enclosure Leakage	0,1mA	0,5mA	0,1mA	0,5mA	0,1mA	0,5mA
Patient Leakage (dc)	0,01mA	0,05mA	0,01mA	0,05mA	0,01mA	0,05mA
Patient Leakage (ac)	0,1mA	0,5mA	0,1mA	0,5mA	0,01mA	0,05mA
Patient Auxiliary Cur. (dc)	0,01mA	0,05mA	0,01mA	0,05mA	0,01mA	0,05mA
Patient Auxiliary Cur. (ac)	0,1mA	0,5mA	0,1mA	0,5mA	0,01mA	0,05mA

Table 2. Limits of leakage currents for different type applied parts (Backes, 2007)

The electrical safety measurements of the medical devices are performed by using an electrical safety analyzer. The analyzer has a software containing the standard of IEC 60601-1 or IEC 62353, or both of them. When the electrical safety measurements are performed, the results are compared to the limit values of the international standards that are loaded in the analyzer and are printed as a report. The report gives the information about whether the device is appropriate to the international standard or not.

For electrical safety, all hospital staff (medical personnel, technical and biomedical staff, hospital management, and even patients) must take on the responsibility. Personnel should not use an extension cord to connect the medical device to the network, should notify the broken / cracked sockets and electric plugs to the relevant technical staff. Any electrical appliance should not be used without a ground connection. Particularly, electrically operated devices and electrical sockets in hospital should be tested in 1 time per year.

9. Performance measurements of medical devices

Patient safety and the quality of medical technology are provided by the performance control of medical devices. Basically, if a sphygmomanometer (noninvasive blood pressure device) does not measure correctly, or an electrocardiography (ECG) does not draw the ECG trace sensitively, it can not be mentioned from the true diagnosis. Wrong diagnosis causes wrong treatment and in this point, some contraventions on patient safety begins. In Figure 11 and 12, the photographs of the performance measurements for the patient monitor, the defibrillator and the anesthesia machine can be seen as an example.

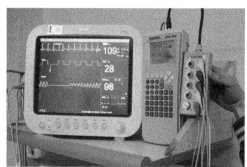

Fig. 11. The performance measurements of the patient monitor and the defibrillator

Performance measurement is the measurement of the accuracy of the medical device or the medical system by using the standard measurement system whose accuracy is known, and is the determination and the record of the deviations. In shortly, by the performance measurements, it is established whether the medical devices are appropriate to the international standards or not, and the problems are also determined if the device is not adequate to the international standards.

The objective of the performance measurement, is to be sure from the accuracy of the medical devices, to minimize the risks and to provide international standardization.

For performance measurements, respectively, the following procedures are observed.

- The medical equipment inventory of the hospital is prepared.
- The medical devices whose performance measurement is needed or not, are determined.
- Performance measurement intervals are determined.
- Performance measurement procedures and measurement forms are prepared.
- Performance measurements are performed in accordance with procedures at the location of the devices and the measurement forms are filled.
- The devices are labeled with the green or red sticker to highlight the performance measurement result.
- Performance measurement certificates are prepared and measurement results are interpreted according to the acceptance criteria in the international standards.
- Performance measurement certificates are archived.

9.1 Medical equipment inventory

All medical devices in a hospital are listed in an inventory. The devices are labeled by using Biomedical Code. Each Biomedical Code should consist of both Universal Medical Device Nomenclature Codes (UMDNS) to define the device and the sign of location. For example the code of ICU 11132 001 means the first defibrilator in intensive care unit. The code of OPR 11132 001 means the first defibrilator in operation room.

9.2 Performance measurement intervals

Performance measurements are not performed for some medical devices that do not have any risk on patients for example nebulizator or ache pump. The technological medical systems including risk are classified as 3 groups. These are:

- High-risk system; They are intensive care, surgical, diagnostic and treatment systems. Their failure or misuse may cause damage to the patient or staff.
- Moderate-risk systems; In case of malfunction, improper use or lack of them, several problems that do not harm the patient or staff seriously may occur in patient care.
- Low-risk systems; The problems about patient care is minimum.

The performance measurement periods are determined by experienced users by considering the device features and the usage conditions. Performance measurement period of devices is calculated with the "Device Management Coefficient" which is used in "Clinical Equipment Management" standards in the "Technology & Safety Management" series developed by the Joint Commission (Fennigkoh et al. 1989). The equation of the "Device Management Coefficient" is shown below.

Device Management = Device Function + Device Risk + Device Preventive Maintenance (2)

    Coefficient   Point   Point   Point

According to the standards, Device Management Coefficient can be maximum of 20 and the devices with 12 or more coefficient are incorporated into the plans of performance measurements (Fennigkoh et al. 1989). If the Device Management Coefficient is greater than 16, the device must be controlled every 6 months.

Device Function Point: It is scored by considering the importance of their function (Table 3)

Point	Device function
10	Life Recovery Devices
9	Surgical and Intensive Care Devices
8	Physical Therapy Devices
7	Surgical and Intensive Care Patient Monitoring Devices
6	Other Physiological Monitors
5	Analytical Laboratory Devices
4	Laboratory Equipment and Supplies
3	Computers
2	Devices that belong to the patients
1	Other devices

Table 3. The points for device function (Fennigkoh et al. 1989)

Device Risk Point: Medical devices have different risks (Table 4) because their absence or failure may cause different problems in patient care.

Point	Device risk
5	Patient death
4	Patient or staff injury
3	Wrong diagnosis or treatment
2	Diagnosis and treatment delays
1	Risk not important

Table 4. The points for device risk (Fennigkoh et al. 1989)

Device Preventive Maintenance Point: Device preventive maintenance points (Table 5) are determined by considering following cases.

- Relevant codes and standards,
- The risks of the device,
- The procedures and test ranges in the device's user and service books,
- The device malfunctions occurred before,
- The service and media properties,
- The usage frequency of the device,
- The device status (old or new, the design problems).

Point	The order of importance
5	Very important
4	Moderately important
3	Less important
2	The least important
1	Minimally important

Table 5. The points for preventive maintenance importance of device (Fennigkoh et al. 1989)

For the calculation of the "Device Management Coefficient", defibrillator device can be given as an example. Firstly, it's function score can be defined as 10 points because it is used for life recovery. The risk score of it can be defined as 5 points because the patient injury or death will be resulted if the device is corrupted or malfunctioning. Finally, the score of the equipment preventive maintenance requirements can be accepted as 4 points by considering the usage frequency of the devices and the risk of breakdown. With the total 19 points, it is decided that the performance measurements of defibrillators should be done every 6 months. In this manner, to determine which devices are taken into the performance measurement system, each medical device is separately handled and the points for all devices are calculated.

9.3 Performance measurement procedures

Measurements are done in accordance to the international standards. Performance measurements are generally performed by using IPM (Inspection and Preventive Maintenance) system procedures that were prepared by Emergency Care Research Institute (ECRI). The principle is to control all parameters of a medical device. The procedures consist of the test steps for both physical and parametrical control. In Table 6, the measurement parameters of some medical devices are listed for illustration.

Fig. 12. The performance measurement of the anesthesia machine

The measurement procedures also consist of acceptable criterias. The results of measurements are interpreted in according to the acceptable limit values. The medical devices whose measurement results are in the range of acceptable values are appropriate to the international standards and are labelled with yellow sticker. It means that the medical device can be used. The medical devices whose measurement results are out of the range of acceptable values are inappropriate to the international standards and are labelled with red sticker. Red sticker means that the medical device should not be used and the device should be controlled by technical personnel or related technical service.

9.4 Performance measurement certificates

The results of performance measurements are given in the calibration certificates. The information to be included in a performance measurement certificate is given below.

- The date of performance measurement,
- The next performance measurement date,
- Situation of the measurement,
- Identity information of the institution performing the performance measurement,
- Information of the test equipment used,
- Environmental conditions at the time of measurement,
- Information obtained during the performance measurement,
- Verification measurements and corrections.

There are some important issues that must be considered during the implementation of the performance measurement method specified in the measurement instructions. They are given below.

Medical Device	Parameters
PATIENT MONITOR	Electrical Safety Tests
	Heart Rate - Fibrillation Arrhythmia
	Respiration Rate - Apnea
	Invasive Blood Pressure
	Noninvasive Blood Pressure
	Cardiac Output
	Oxygen Saturation Concentration
	Temperature
ANESTHESIA MACHINE	Electrical Safety Tests
	Gas Concentration (Isoflourane,......)
	Tidal Volume
	Minute Volume
	Inspiration Time
	Expiration Time
	I/E Ratio
	PEEP
	Mean and Maximum Pressure
	Oxygen Concentration
ELECTROCARDIOGRAPHY	Electrical Safety Tests
	Linearity Test
	Sensitivity Test
	1mV Pulse Test
	Paper Speed Test
DEFIBRILLATORS	Electrical Safety Tests
	Energy Measurement
	Charge Time Test
	Battery Test
	ECG and Arrhythmia Simulation

Table 6. The measurement parameters of some medical devices

- Performance measurements should be done by trained personnel.
- The international traceability of the test equipment must be ensured.
- The error rate used in the measurement process should be known.
- The accuracy of the test equipment must be very small than the accuracy of the device which is tested.
- Technical specifications of the device must be determined.
- The environmental conditions such as heat, moisture, vibration, dust, electromagnetic field strength, temperature change rate etc. should be checked during the performance measurements.

- Measuring range must be identified and measurements must be screened in this range,
- The repeatability of the measurements must be provided.

10. Conclusion

As a result, it is seen that one of the important components of patient safety is medical device safety. By increasing medical device safety, patient safety increases.

By taking the necessary measures related to all considerations mentioned in the above sections;

- Prevention of medical accidents that may occur in the health establishment,
- Reducing the risk of infection,
- Ensuring the accuracy of medical devices,
- Routine controlling of medical devices,
- Increasing of both employee and patient satisfaction, are possible.

Therefore, the "patient safety" is under control as a result of good management of medical technology.

11. References

AAMI, (1995). *Sterilization Part 1. Sterilization of Healthcare Facilities. Sterilization Part 2. Hospital Eqipment and Industrial Process Control*, Association for the Advancement of Medical Instrumentation, USA

AMN Healthcare, (2008). Fire Starter. in Action Steps, (2008)

ANSI AAMI ST79, (2006). *Comprehensive Guide to Steam Sterilization and Sterility Assurance in Health Care Facilities.* Association for the Advancement of Medical Instrumentation, ISBN 1-57020-256-7, USA

Avitall, B., Khan, M., Krum, D., Jazayeri, M., & Hare, J. (1993). Repeated use of ablation catheters: A prospective study. *Journal of American College of Cardiology*, Vol.22, No.5, (1993), pp. 1367-1372

Backes, J. (2007). *A Practical Guide to IEC 60601-1*. Rigel Medical, United Kingdom, (2007)

Barbosa, A.T.R., Iaione, F., & Spalding, L.E.S. (2010). In a hospital: an electrical safety and information system. *32nd Annual International Conference of the IEEE EMBS*, (2010), pp. 4427-4430

Bassen, H.I. (1998). COMAR technical information statement: Radiofrequency interference with medical devices. *IEEE Engineering in Medicine and Electronics Engineers*, Vol.17, No.3, (1998), pp. 111-114

Brennan, T.A., Leape L.L., Laird N.M., & Hebert, L. (1991). Incidence of adverse events and negligence in hospitalized patients: Results of the Harward Medical Practice Study. *N Engl J Med.*, Vol.324, No.6, (1991), pp. 370-376

Buchdid Amarente, J.M., Toscana C.M., Pearson M.L., Roth, V., & Jarvis W.R. (2008). Reprocessing and reuse of single-use medical devices used during hemodynamic procedures in Brazil: A widespread and largely overlooked problem. *Infect Control Hospital Epidemiol*, Vol.29, No.9, (2008), pp. 854-858

Carranza, N., Febles, V., Hernandez, J.A., Bardasano, J.L., & Monteagudo, J.L. (2011). Patient safety and electromagnetic protection: A review. *Health Physics*, Vol.100, No.5, (2011), pp. 530-541

Carol, R. (2003). FDA works to reduce preventable medical device injuries. *FDA Consumer*, Vol. 37 (4), July (2003)

Chakrabartty, A., Panda, R. (2010). Criticality of electrical safety for medical devices. *Proceedings of 2010 International Conference on Systems in Medicine and Biology*, (2010), pp. 212-216

Day, P. (2004). What is the evidence on the safety and effectiveness of the reuse of medical devices labelled as single-use only? *New Zealand Health Technology Assessment*, Vol.53, (2004)

Dubois, V. (2002). Sterilization techniques for medical instrumentation in health care facilities. Vide-Science Technique et Applications, Vol.57, (2002), pp. 85-91

Emergency Care Research Institute (2006). FDA issues statement on reuse of single-use devices. *FDA Normal Priority Medical Device Alert*, (2006).

Fennigkoh, L., Smith, B. (1989). Clinical equipment management. *JCAHO Plant, Technology & Safely Management Series (2)*, (1989), pp. 5-14

FDA (1997). A primer on medical device interactions with magnetic resonance imaging systems. *FDA's Good Guidance Practices*, (February 1997)

FDA (2000). FDA releases final guidance on the reprocessing and reuse of single-use devices. *Medical Device Reporting*, No. 31, (2000)

Gilligan, P., Somerville, S., & Ennis, J.T. (2000). GSM cell phones can interfere with ionizing radiation dose monitoring equipment. *The British Journal of Radiology*, Vol.73, (2000), pp. 994-998

Hailey, D., Jacobs, P.D., Ries, N.M., & Polisena, J. (2008). Reuse of single use medical devices in Canada: Clinical and economic outcomes, legal and ethical issues, and current hospital practice. *International Journal of Technology Assessment in Health Care*, Vol.24, No.4, (2008), pp. 430-436

Hans, N. & Kapadia F.N. (2008). Effects of mobile phone use on specific intensive care unit devices. *Indian Journal of Critical Care Medicine*, Vol.12, No.4, (2008), pp. 170-173

Henney, J.E. (2000). FDA's proposed strategy on reuse of single use devices. *Journal of the American Medical Association*, Vol.283, No.1, (January 2000), pp. 46

Hijazi, R. (2011). The impact of medical devices on patient health. *Journal of Clinical Engineering*, (July/September 2011), pp. 105-108

IAEA (1996). *International Basic Safety Standards for Protection against Ionizing Radiation and for the Safety of Radiation Sources*. IAEA Safety Series 15, ISBN 92-0-104295-7, Vienna, Austria

ICRP (1991). 1990 Recommendations of the international commission on radiological protection. ICRP Publication 60, *Annals of the ICRP*, Vol.21, No.1-3, (1991)

IEC 60601-1 (2005). *Medical Electrical Equipment-General Requirements of Safety*. International Electrotechnical Commission (IEC), (2005)

IEC 62353 (2007). *Medical Electrical Equipment-Recurrent Test and Test after Repair of Medical Electrical Equipment*. International Electrotechnical Commission (IEC), (2007)

ISO 14644-1:1999 (1999). *Cleanrooms and associated controlled environments- Part 1: Classification of air cleanliness*, International Standardization for Organization, (1999)

Kelkar, U., Bal, A.M., & Kulkarni, S. (2004). Monitoring of steam sterilization process by biologic indicators-a necessary surveillance tool. American *Journal of Infection Control*, (2004), pp. 512-513

Koh, A. (2005). Current practices and problems in the reuse of endoscopic single-use devices in Japan. *American Journal of Infection Control*, Vol.33, No.5, (2005), pp. 156

Lawrentschuk, N., & Bolton, D.M. (2004). Mobile phone interference with medical equipment and its clinical relevance: a systematic review. *The Medical Journal of Australia*, Vol.181, No.3, (2004), pp. 145-149

McDermott, C. (2010). Regulations that impact disinfection and sterilization processes: ALPHA Patrol: Keeping health care safe for everyone. *Perioperative Nursing Clinics*, Vol.5, (2010), pp. 347-353

Nobel, J.J. (1996). Medical device accident reporting: does it improve patient safety? *Stud Health Technol Inform*, Vol.28, (1996), pp. 29-35

Northrup, S.J. (2000). Reprocessing single-use devices: An undue risk. *Medical Device Manufacturers Association*, (2000)

Osman, M.A., Todorova, A., & Samra A.H. (1996). Electrical safety in medical institutions-Neutral electric potentiality system in hospitals. *Proceedings of IEEE Southeastcon 97 Conference on Engineering the New Century*, (1996), pp. 304-306

Quirk, M. (2002). Most Us hospitals avoid reuse of single-use devices. *The LANCET Infectious Diseases*, Vol.2, (December 2002), pp. 714

PMDA (2006). *Effects on implantable medical devices /cardiac pacemakaers and cardioverter defibrillators) by new system mobile phone terminals*. PMDA Pharmaceuticals and Medical Devices Safety Information, No.226, (2006), Tokyo, Japan

Popp, W., Rasslan, O., Unahalekhaka, A., & Brenner, P. (2010). What is the use? An international look at reuse of single-use medical devices. *International Journal of Hygiene and Environmental Health*, Vol.213, (2010), pp. 302-307

Pressly, N. (2000). FDA warns about EMI risk with telemetry systems. *FDA User Facility Reporting Bulletin*, (2000)

Rados, C. (2004). FDA works to reduce preventable medical device injuries. *FDA Consum.* No. 37(4), (2004)

Rice, L.R.H., Albertson, L.V.E., Anderson, P., & Day, M. (2009). Reuse of single –use critical medical devices. *The Official Journal of The Society of Gastroenterology Nurses and Associates,* Vol.32, No.3, (2009), pp. 228-229

Rizzo, J., Bernstein, D., & Gress, F. (2000). A performance, safety and cost comparison of reusable and disposable endoscopic biopsy forceps: a prospective, randomized trial. *Gastrointestinal Endoscopy*, Vol. 51, No. 3, (2000), pp. 257-261

Rutala, W.A., & Weber, D.J. (2004). Disinfection and sterilization in health care facilities: What clinicians need to know. *Healthcare Epidemiology*, Vol.39, (2004), pp. 702-709

Sawyer, D. (1997). An introduction to human factors in medical devices. *U.S. Department of Health and Human Services, Food and Drug Administration*, (1997).

Selvey, D. (2001). Medical Device Reprocessing Is it good for your organization? *The FDA Guidance Document*, (2001)

Sezdi, M. (2009a). The effects of medical device accidents and user errors to patient safety. *Proceedings of 1st. International Conference on Patient Rights*, (November, 2009)

Sezdi, M. (2009b). Is it possible that the death reason of a catheterized patient is the leakage current? *Proceedings of 1st. International Conference on Patient Rights*, (November, 2009)

Sezdi, M. (2009c). Particle measurement errors in intensive care units. *Proceedings of 3th International Conference on Quality in Healthcare Accreditation and Patient Safety, (2009)*

Tan, K.S., Hinberg, I., & Wadhwani, J. (2001). Electromagnetic interference in medical devices: Health Canada's past and current perspectives and activities. Journal of IEEE Electromagnetic Compatibility, Vol. 2, (2001), pp. 1283-1288

Webster, J.G. (1992). *Medical Instrumentation Application and Design* (2nd edition), Houghton Mifflin Company, ISBN 0-395-59492-8, Boston, USA

Yoleri, G. (2011). *Technical and Biological Performance Tests for Autoclave Units* (Master Thesis), Istanbul University, Istanbul, Turkey

Zimerman, L.I., Cenci, F., & Streck E.E. (2003). The effects of multiple reprocessing of radiofrequeny ablation catheters on their electrical integrity. *Progress in Biomedical Research*, Vol.8, No.2, (2003), pp. 116-118

7

Therapeutic Lasers and Skin Welding Applications

Haşim Özgür Tabakoğlu[1] and Ayşen Gürkan Özer[2]
[1]Fatih University, Institute of Biomedical Engineering, İstanbul,
[2]İstanbul Technical University, Faculty of Art and Sciences, İstanbul,
Turkey

1. Introduction

1.1 Laser physics

Laser systems used in therapeutic medical field are specialized in terms of energy production and optical delivery mechanisms. It is favorable that medical lasers should have compact design especially important for a surgeon who is working in operation room where bulky medical devices occupy wide spaces. With the developing technology, especially in diode electronics, lasers of dimensions as small as i.e. 20x20x20 cm³ can be produced. Laser physics can be overviewed elsewhere written in any physics or biomedical books. Here we will only mention about important mechanisms and general knowledge.

1.2 Laser light

Since the word LASER is an acronym standing for Light Amplification by the Stimulated Emission of Radiation, the term optical amplifier can be considered as a concise laser device. In the atomic structure, electrons being in the quantized energy levels can only move with respect to the Quantum Physics rules. As seen on Figure 1, motions of the electrons, in other words electronic transitions cause the three main phenomena producing laser light. These are spontaneous absorption, spontaneous emission, and stimulated emission that determine the main features of the laser light.

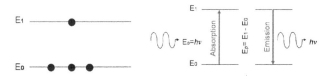

Fig. 1. Quantized energy levels, absorption and emission phenomena.

Spontaneous absorption, or simply absorption is the increase in an electron's energy level by capturing a photon's energy. The captured energy here is the key process meaning the exact difference between the energy levels as the word "exact" means quantized here. After the absorption, the atom changes its state from stable ground energy level E_0 to unstable excited energy level E_1, Figure 2.

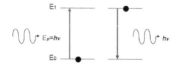

Fig. 2. Schematic drawing of absorption.

Becoming unstable, the atom needs to be stable again so it emits a photon to become stable again. This is called emission. During emission, the captured photon is released and luminescence takes place. If the emission occurs spontaneously, the process is named *spontaneous emission*, Figure 3; but we are not interested in this kind of emission in this text.

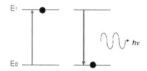

Fig. 3. Schematic drawing of spontaneous emission.

The last key process, *stimulated emission* occurs when another photon containing exactly the same energy between E_0 and E_1 hits, or "stimulates" the electron being at the excited energy level E_1, Figure 4. After this step, there are two photons identically the same with each other. Their directions, wavelengths, and phases are the same. These two photons which form the unique laser light are called coherent.

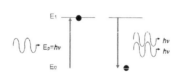

Fig. 4. Schematic drawing of stimulated emission.

In order to keep laser light under control, its production requires some specifications which are main parts of a laser devise: resonator, medium, and pump.

The *resonator*, or cavity, can be almost considered as a closed box. There are two mirrors generally located face to face. One of them, M_1, is 100% reflective, and the other one, M_2, is 98-99% reflective. M_2 has a hole to allow the photons reflected from M_1 to escape and the output is named as laser light.

The *medium* is the substance to be used for generating photons from its electrons indeed. It is situated in the resonator and determines the wavelength. Since the laser medium can be solid, liquid, or gas, the wavelength range can be easily selected from UV to IR.

The *pump* is the energy source which starts the population inversion and it can be in several ways such as electrical, optical, thermal, or chemical. The process population inversion is the most important step of laser light production and can be basically summarized as to make the number of atoms in the excited state more than the number of atoms in the ground state.

In short, when the pump, resonator and medium come together, the pump starts population inversion and photons released by spontaneous emission cause much more photons released by stimulated emission. Thus photons proliferate and start to move back and forth between the mirrors M_1 and M_2 until reaching certain intensity then pure laser light quits the resonator that becomes the laser device.

1.3 Interaction mechanisms

Interaction of radiation with matter is summarized above, but, when the matter is tissue specifically, the interaction mechanisms become important. There are mainly five different laser-tissue interaction mechanisms observed on tissues. These are *photochemical, photothermal, photoablation, plasma-induced ablation, and photodisruption,* see Figure 5. Main concern of these mechanisms is the deposition of delivered laser energy which is determined by *laser parameters* such as wavelength, spot size, exposure time, pulse duration, repetition rate, etc.; *optical tissue properties* such as anisotropy factor, absorption and scattering coefficients; and, *thermal tissue properties* such as heat capacity, heat conduction, tissue density, etc. As a result of these parameters, there is one more important parameter called thermal relaxation time.

As seen on Figure 5, if the exposure time lasts long, power density decreases; if the power density increases, exposure time decreases, but both are almost always kept in between 1 J/cm^2 and 1000 J/cm^2.

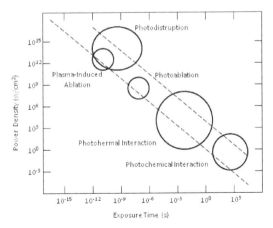

Fig. 5. Laser-tissue interaction types. (Modified from Niemz.) [1]

Photochemical interactions occur at very low power densities with relatively long exposure time, from seconds to days. One of the best known applications is a kind of tumor treatment, Photo-Dynamic Therapy (PDT), which uses a catalytic photosensitizer to destroy the tumor.

Photothermal interactions are very common medical applications and occur at relatively low power densities. Depending on the exposure time, four major effects are observed on the tissue; *hyperthermia, coagulation, carbonization, and vaporization* which determine the type of treatment whether it is diagnostic or therapeutic. Energy of the photons is absorbed by the tissue and transformed into heat, and, depending on heat propagation and deposition in tissues, the photothermal effects originate. For example at room temperature (25°C), a tissue normally can heat up to 37°C without any damage and after this point keeps heating with a mild change in color. This change is named as *hyperthermia* and the tissue can still heal. After 50°C, enzyme activity decreases, cells cannot move and the irreversible damage *coagulation* (a kind of sclerosis) occurs at 60°C. Actually, the temperature window 60 – 65°C are very sensitive for the collagen structures covering tissues, because this is the basic principle of **tissue welding**. If the coagulated tissue continues to heat, the water molecules being in the tissue constituents start to vaporize (*vaporization*) and pressure starts to appear at the inner structures at 100°C. Consequently, tissues start to leave the application area, thus, thermal ablation happens with the thermal-mechanical effects. Since the expansion of the pressure causes tissues to tear and burst, this type of ablation is completely different from the photoablation. As a result of this, damages at the vicinity and the ablation crater on the application surface form. Beyond 200°C, *carbonization* starts and is used to burn some kinds of tumors on the surface. When the temperature reaches 300°C, *melting* is observed. All these effects can be seen on Figure 6 [2].

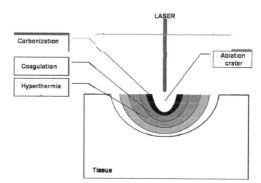

Fig. 6. Photothermal effects observed inside biological tissue (Modified from Niemz) [1].

Photoablation needs high energy photons from UV region of the electromagnetic spectrum, so it occurs at relatively high power densities with short exposure times. Since the main principle here is directly breaking the molecular bonds, edges of the ablation volume are very smooth, clean, and regular. This process is mostly used for corneal shaping to achieve vision correction in Ophthalmology.

Plasma-induced ablation is the ablation by ionizing plasma formation. Seen on the Figure x, when the power density reaches 1012 -1016 W/cm², an electric field larger than the average Coulombic attraction force between the electrons and the nuclei occurs and creates a very large free electron density, the plasma, causing dielectric breakdown of the tissue. This is called optical breakdown and the plasma absorbs very well from the UV to IR and creating very clean ablation allows many treatments especially used in Dentistry.

Photodisruption is achieved by plasma formation again and observed on the soft tissues or fluids. Since the formation of plasma needs higher energies, shock waves appear in the tissue and 'disrupt' the tissue structure with a mechanical effect. Because of both the plasma-induced ablation and photodisruption shares a big common area on the graphic, Figure x, it is difficult to distinguish them from each other [1-3].

2. Skin welding applications

Improved wound healing strategies tried to be investigated by biomedical research groups. Sealing a cut with beam of light to its immaculate state like in science fiction movies seems far to be accomplished that is like earlier men's desire to fly. Will humanity eventually be able to come closer the supernatural healing? The efficacy of wound closure strategies will need to be tested step by step in scientific manner.

Closure or anastomosis of tissue by laser welding and soldering occur due to two phenomena: First one is physical that is thermal and optical interactions of laser with tissue and the second one is biological that is modifications of connective tissue proteins during heating process and reassembly. Since the first application in 1964 on small blood vessels with Nd:YAG laser by Yahr and Strully [4], different laser systems have been tried at different tissue types, have transferred from laboratory testing to approved clinical practice [5]. The first attempt came from Schober *et al* [6] about finding molecular level changes of laser irradiated tissue but process that is change in collagen fiber bundles with interdigitation of altered individual tropocollagen units, denaturation and cross-linking of tissue proteins seems to be structural mechanism of the welding effect, could not be understood thoroughly [7-14].

Natural tissue repair is considered as three phases and followed by histological staining techniques: (I) cellular migration and inflammation for the very first days; (II) fibroblasts proliferation for 2 to 4 weeks, with new collagen synthesis; and (III) remodeling from 1 month to 1 year. This process includes collagen cross-linking and active collagen turnover [15], in general. Healing phase further divided into phases of coagulation, inflammation, matrix synthesis and deposition, angiogenesis, fibroplasia, epithelialization, contraction, and remodeling, [16-18] in detail. Fibroblast cells, as being the principal source of collagen and wound connective tissue, begin to synthesize and secrete measurable amounts of extracellular collagen on the second or third day after wounding. Polymerization of collagen fibers occurs in extracellular wound environment where monomers of collagen are secreted. Tensile strength of collagen fibers increases as increase in covalent cross-linking among fibers. At 1 week following wounding, immature collagen fibers become histologically apparent in the wound [16, 19-21].

Requirement of mechanical assembly in any skin cuts or surgical incisions was carried out by using conventional mechanical closure techniques for tissue bonding (sutures, staples, and adhesives) are highly reliable procedures that have proven themselves over the years to be good clinical practice [22-28]. These methods are favored due to reliability, cost-effectiveness and suitability of any type of tissue. The primary function of the suture is to maintain tissue approximation during healing. Sutures placed in the dermal layer provide tensile strength, and control tension for the outer layer [22, 29, 30]. In the selection of a suture, a patient health status, age, weight, and the presence or absence of infection are as important as the biomechanical properties of the suture, individual wound characteristics, anatomic location, and a surgeon's personal preference and experience in handling a suture material. Fundamental intention to any surgeon is choosing a method of closure that affords a technically easy and efficient procedure, secure closure, minimal pain and scaring. However, these conventional fasteners cause tissue injury due to their mechanical penetration. Foreign body reaction is given rise because of nature of materials used [31, 32]. Tissue injury and foreign-body reaction can induce inflammation, granuloma formation, scarring, and stenosis [33]. Sutures may not be suitable for microsurgical or minimally invasive endoscopic applications [34]. Precision of alignment is so hard for staples and clips due to relatively large force requirements for positioning. Most importantly, none of these fasteners produces a watertight seal over the repair [35]. Besides the mechanical closure methods, FDA approved biological and non-biological topical skin adhesive chemicals such as fibrin glue and cyanoacrylate based compounds have been used in skin edges of wounds from surgical incisions, and simple, trauma-induced lacerations [27, 36-38]. Adhesives may be used together with dermal stitches although it cannot be used in place of that technique. Chemical bonding of tissue has contraindications over hypersensitive patients to cyanoacrylate or formaldehyde, wound with evidence of infection, mucosal surfaces [39, 40]. They are toxic to the tissues, not absorbed in the normal wound healing process, and cause foreign-body granulomas and allergic reactions.

Increased instant wound strength,fluid-tight closure, low level risk of infection are the advantages of laser tissue closure over conventional methods [35, 41-44]. On the other hand, thermal damage to tissue was the biggest handicap of this technique. This is an unwanted event opposed to the objective of laser tissue repair methods that is to obtain coagulation of a desired volume of tissue with minimal effects in the surrounding tissue. Endpoint decision for the tissue apposition and poor reproducibility are the other negative sides of application [45]. Optimization of laser tissue closure is quite hard due to high number of tissue optical and thermal parameters as well as laser parameters. Tissue optical and thermal parameters cannot be changed but by addition of any external agent laser light can be localized in a specific tissue region. On the other hand, laser parameters such as wavelength, fluence or irradiance, pulse duration, repetition rate, irradiation time and spot size can be adjusted to get successful tissue closure [5, 45, 48]. In the following sections optimization studies of any of these parameters will be introduced.

2.1 Tissue closure and anastomosis by medical lasers

Two types of tissue welding can be defined: Laser Welding and Laser Soldering. In order to hold together, target tissue is irradiated with laser application in laser welding method. In the second type of application, use of soldering materials and wavelength specific absorptive dyes can enhance tissue sealing with the selective heating of target tissue.

Photothermal interaction occurs in a way that, laser (photon) energy, applied directly to the tissue surface is converted to the heat energy by molecular vibration of tissue chromophores such as water, melanin, hemoglobin by absorption (Figure-7). The rate of heat generation depends on the rate of absorption of photons within the tissue [47]. Scattered light that is absorbed may cause heating outside the laser beam. Increase in temperature to a certain degrees causes structural changes (interdigitation) in tissue proteins such as collagen and fibrinogen, in a way that they can bond each other in their open sites at the cooling phase. can cause irreversible damage of the proteins of the tissue (Table-1). Thermal response of laser irradiated tissue have been examined in detail and modeled by Welch *et al* [47-51].

Laser tissue welding is photothermal process

INTERACTION OF RADIATION WITH MATTER

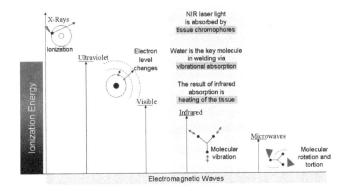

Fig. 7. Heating of tissue by infrared mediated molecular vibration.

Temperature	Molecular and Tissue Reactions
42-45 °C	Hyperthermia leading to protein structural changes hydrogen bond breaking. retraction
45-50 °C	Enzyme inactivation changes in membrane permeabilization, oedema
50-60 °C	Coagulation, protein denaturation
~ 80 °C	Collagen denaturation
80-100 °C	Dehydration
>100 °C	Boiling, steaming
100-300 °C	Vaporization, tissue ablation
>300 °C	Carbonization

Table 1. Laser Tissue Interactions: Photothermal Effects [44]

Wide range of medical lasers has been used for laser tissue welding. Lasers mainly operate in infrared region such as carbon dioxide (CO_2), THC:YAG, Ho:YAG, Tm:YAG, and Nd:YAG, and gallium aluminum arsenide diode (GaAlAs) lasers are suitable for tissue welding [52-55]. The CO_2 laser has been preffered repairing of thin tissues because of relatively short penetration depth (shorter than 20 μm) in tissue (Table-2). On the other hand, in thick tissues, high irradiances or long exposure times are necessary for sealing of target tissue [47]. Those types of applications cause undesired consequences such as carbonization of tissue and the conduction of heat from the initial absorption zone to surrounding tissue [45].

Laser	λ (nm)	δ_{water} [53, 54]	δ_{tissue} [54, 55]
Argon ion	488	23 m	0.8 mm
KTP/Nd:YAG	532	16 m	1.1 mm
HeNe	632.8	4.8 m	3.5 mm
GaAlAs	780	60 cm	7 mm
	820	46 cm	8 mm
	870	25 cm	7 mm
Nd:YAG	1064	4 cm	4 mm
Nd:YAP	1080	5 cm	4 mm
Ho:YAG	2100	0.2 mm	1 mm
Er:YAG	2940	1 μm	<1 μm
CO_2	10600	10 μm	20 μm

Table 2. Penetration Depths of Some Medical Lasers in Water and in Tissue

Temperature controlled laser systems have been developed to get rid of over-heating consequently thermal injury to the surrounding tissue. Researchers applied radiometric methods to determine the temperature of the surface of the heated cut and used computer assisted temperature feedback control to maintain tissue surface temperature at some desired level. The Applied Physics group at the Tel Aviv University developed a laser fiberoptic system for monitoring and controlling the temperature [59, 60] for CO_2 laser welding and/or soldering. Polycrystalline silver halides (AgClxBr1-x), which are highly transparent in the mid-infrared spectral region (3 to 30 μm) used in optical fibers transmits emitting infrared radiation from surface of laser application spot to radiometric detector which is connected to the personal computer. The personal computer controls the power emitted by the laser, and thus temperature. The system has been tried in various kinds of tissues. Skin tissue closure in rabbits, wound healing pattern was observed on 3rd, 7th, 14th and 28th post operative days of healing period by Simhon et al. [61]. Laser application along incision was done spot by spot. The mean laser power density at the heated spots was 7 W/cm2. The spot size was 3 mm in diameter on tissue surface. Histology examinations showed thermal injury, as basophilic coloring, necrosis or carbonization residues, the presence of surface albumin residues, and degree of re-epithelialization. Laser skin closure with thermal control was found to be more successful to other wound-closure modes. Thermal damage was minimized by the use of a temperature controlled fiber optic laser-soldering system. Tensile test performed as a complimentary for this study [62] revealed immediate tensile strength values that were similar to those obtained with cyanoacrylate glues. Moreover, laser-soldered incisions exhibited stronger long-term tensile strength and presented better wound healing behaviour, akin sutured incisions.

In vivo study was also performed in porcine model by Simhon etal. [63] aiming toward clinical trials by evaluating the efficacy, reproducibility, and safety of the temperature controlled laser system procedure in pigs. 500 mW CO_2 laser power was applied to the skin tissue. Therefore, the optimal soldering temperature was found to be within the range of 65°C to 70°C. At this range, successive tensile strength and tight sealing of the wound was accomplished.

The system has been tried also in soft tissue welding in different animals.

It was found that laser welded tissue has been much less scar tissue compared to control group (standard suture) at the end of 30-day inspection period [64]. Rat intestine argon laser (operates at multiple visible wavelengths from 488 nm to 515 nm, irradiance was 28 W/cm2) tissue fusion was carried out by Çilesiz *et al.* [65] found that laser-assisted intestinal anastomoses provided an immediate fluid-tight closure compared to the suture control anastomoses. Less thermal hazard were inspected when laser irradiation was controlled by thermal feedback. At 3 weeks, all anastomoses were mechanically as strong as intact intestine, although bursting/leaking pressure tests of laser-assisted anastomoses without thermal feedback control (TFC) tended to be lower than all sutured anastomoses (control) group and laser tissue welding with TFC group. Experimental findings indicated that; although welds were initially weak, they were sufficiently stabilized to resist spontaneous rupture in surviving animals. In similar study performed by Çilesiz *et al.*, Ho:YAG (2.09 µm, irradiance was 16 W/cm²) laser was used to anastomose rat intestine with and without TFC at 90°C [66].

Simple and easy-to-apply methods are demanded in minimally invasive surgery. In a study performed by Spector *et al.* [67] small bowel harvested from 6-month-old pigs was successfully anastomosed by using albumin stent heated up by 828 nm GaAs semiconductor laser diode which emission was delivered to the tissue by 600 µm silica fiber. Experimental setup was established such that the laser delivery fiber and the radiometer fiber was positioned 8 mm from the tissue, producing a 3.4 mm diameter spot with a maximal power density of 30 W/cm². Albumin stent laser anastomosis of bowel sustained statistically significant higher bursting pressures than those done by sutures. The addition of solder materials (blood, fibrinogen, and albumin) and wavelength specific chromophore help to strengthen the wound, to maintain edge alignment and to focus energy to the target region. Furthermore, due to the increased absorption characteristics of the dyed tissue or solder, low-level laser powers may be used to achieve the required effect, increasing the safety of the technique. The combination of serum albumin and ICG dye with an 800-nm diode laser has been tried in numerous of applications in order to   nd optimal solder composition [67-73].

One of the most comprehensive studies was performed by McNally *et al* [74] to investigate the relative importance of these parameters to laser tissue soldering. Twenty five different combinations of laser irradiance and exposure time were used. The effect of changing bovine serum albumin concentration and indocyanine green dye concentration of the protein solder on the tensile strength of the resulting bonds was investigated on *ex vivo* aorta specimens irradiated with 808 nm diode laser delivered 400 µm-core silica fiber. Spot size at the protein solder surface was 1 mm. Mechanical test and electron microscopy analysis were performed in order to investigate bond stability in hydtarion point of view. It was observed that the failure mechanisms were of two kinds. The liquid protein solder broke into two halves, but each remained attached to the tissue. On the other hand, the solid protein solder, remained intact but detached from the tissue. The overall tensile strength of repairs formed using the solid protein solder were significantly higher than the strength of the liquid protein solder repairs was noted. Although achievements have come to a considerable point, technical problems should be overcome. Three disadvantages of the solid protein solder have been indicated [75]. First, a non-uniform solder coagulant generally results in because of temperature gradient over depth of the solder. Second, the protein solder is soluble. Third, the solid protein solder is not flexible enough to fit different tissue geometries. Specially designed artificial solder- doped PGA membranes were tried for laser tissue soldering by McNally *et al.* and Hodges *et al.* [69, 75]. The albumin protein solder and the polymer materials are biodegradable. This brings about the minimal foreign body reaction and infection. PLGA based synthetic materials were prepared. Average irradiances of 5.7, 11.3, and 17.0 W/cm² delivered to the solder surface [75]. Results showed that the solder-doped polymer membranes improved repair strength as well as exibility during application over previous published results with albumin protein solders. Moreover tissue apposition can be established by rehydration of the solder-doped polymer membranes upon application.

Lauto *et al.* examined *in vitro* and *in vivo* tissue repair with chitosan adhesive [76]. Chitosan gels have been shown to induce no thermal damage to tissue and to produce better sealing [77]. Chitosan adhesives provide manipulation of tissue without breaking or tearing. The adhesive can be hold when manipulated with forceps and appeared to be well suited for tissue repair A review paper about different tissue reconstruction strategies employing adhesive biomaterials currently used in surgical and experimental procedures was written by Lauto PhD. concluded that each of these adhesives has an optimal clinical application depending on its physical-chemical characteristics (adhesiveness, fluid or solid consistency, chemical or light activation for example). It is thus very likely that different bioglues will be adopted by surgeons for their specific operative procedures, instead of a single product [78].

3. Materials and methods

3.1 Surgical applications

All experiments were performed under authorization of Institutional Animal Research and Care Ethic Committee at Boğaziçi University (BUHAYDEK, Date:1/12/2005, No:2005-6). Male Wistar Rats, randomly selected, 7-9 months old, weighing 290-320 g, from Psychobiology Laboratory of Boğaziçi University were used in all experiments mentioned in this study. Rats were housed in plastic cages and maintained on a 12-h-light/12-h-dark cycle in a temperature-controlled vivarium (22±2°C). Food and water were available *ad libitum*.

Surgery

Wistar rats were anesthetized with ketamine (10% ketamidor, RichterPharma, AG, Wels, Austria) by intraperitoneal injection (1.65 ml/kg). Hair at the site of application was shaved. Antiseptic Poviiodex Scrub (Kim-Pa İlaç Lab. Inc. Hadımköy) was applied topically to prevent infection and desiccation of the wounds. Three pairs of 1-cm-long full-thickness incisions (over muscular layer), bilateral and parallel to the spinal cord, were done. Incisions were implemented with sterile No.11 surgical blade (Tontarra Medizintechnik GMBH, Germany). In case of bleeding, incision site was compressed and any blood remnant was cleaned from top of skin in order to prevent probable light absorption by blood remnant on the surface of skin rather than skin tissue itself. Incision was closed with near infrared lasers (either 809 nm, 980 nm or 1070 nm) or suturing that was used as control group if included in any of experiment.

Post Surgery

After closure, Thiocilline (Abdi İbrahim İlaç Inc., İstanbul, Turkey) was applied on each incision to inhibit any kind of microbial reactions on incisions. The length and the thickness of the incisions were checked via digital caliper (Mitutoyo, UK). No food or water was given 24-h post-surgery. Investigation of healing was performed through 4-day period for predosimetry studies and 21-day for comparative studies. 1st, 4th, 7th, 14th and 21st days were selected as control days during healing period. On these particular days, skin samples, closed either by NIR lasers or suture were collected for histology or tensile testing. During healing period, each rat was kept in separate cage until histology examination or tensile test on particular post operative days. At the end of the any experiment mentioned rats were anesthetized first and then killed (cervical dislocation).

3.2 Closure methods: Laser systems and suturing

809-nm diode laser system

The 809-nm Diode Laser System have been designed and improved in Boğaziçi University, Institute of Biomedical Engineering, Biophotonics Laboratory [79]. The system is a computer controlled high power 809-nm laser (10 W output power at 35 A applied current). User interface was developed in C programming language communicates with the controller unit of the diode laser system to set the operating parameters of the diode laser module [80]. Laser output ber was coupled with 400 nm fiber via FC (Fixed Connection) connector to transfer laser energy to the target tissue part.

980-nm diode laser system

The 980-nm diode (OPC-D010-980-FCPS, Opto Power, Tuscon, AZ, USA) class IV laser was controlled by a microcontroller-based controller instrument, which was designed and manufactured by our group at Biophotonics Laboratory at Institute of Biomedical Engineering, Boğaziçi University [81]. The laser unit can also be controlled using OPC remote control module provided with the system. Either control device attaches to the laser system via the 25-pin D-sub connector on the front panel. In general description, high power diode laser system has 10W maximum optical output. The system provides output through a 1 meter fiber-optic cable. Laser was delivered to the target tissue with a 400µm optical fiber (Spindler-Hoyer, Gottingen, Germany) via subMiniature version A (SMA) connection. The parameters (power, exposure time, number of cycles, and on-off duration of pulses) of the laser were controlled with interface developed in LabView software (V 6.1).

1070-nm Ytterbium fiber laser

Device produced by IPG Laser GmbH and produces 20 Watts of optical power at a wavelength of 1070 nm (YLM-20-SC Series, Germany). In order to set number of pulses, power and exposure time, an external microcontroller based controller (Teknofil Inc., Istanbul, Turkey) was designed and used in experiments

Laser delivery

Laser closure was performed that six spots were applied by a handpiece made up of plexiglass material at a distance of 2 mm above 1-cm long incision (Figure 8). Each spot had a 2 mm diameter and they slightly overlapped onto each other. Five seconds cooling time interval was given between spots. Spot size (0.0314 cm^2) was checked with a detection card (VRC4, Thorlabs, NJ, USA) and power of the laser was checked with a powermeter (Newport 1918-C, CA, USA) before each application.

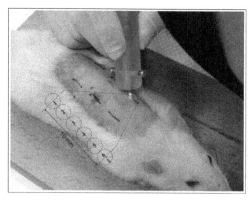

Fig. 8. 6 laser spots along 1cm incision. 2 mm spot size was obtained from optic fiber that was placed 2 mm above skin surface.

Suturing

A single, interrupted suture USP-3/0 metric silk (Doğsan Tıbbi Malzeme San. A.Ş. Trabzon-Turkey) was placed in the middle of the 1 cm long incisions. Distance between needle entrances to the tissue was about 10 mm.

3.3 Histology examinations

Incision sites were removed one by one, in rectangular shape (1x2cm) from rat's dorsal skin and stored in tissue fixative (10%PBS formaldehyde) until tissue processing that was automatically performed by tissue processing machine (LeicaTP-1020, Germany). Paraffin embedding and tissue sectioning were performed. Hematoxylin and Eosin (H&E) staining was used for histological examinations. This stain gives idea about general tissue structure. Hematoxylin, a basic dye, stains nuclei blue due to an affinity to nucleic acids in the cell nucleus; eosin, an acidic dye, stains the cytoplasm, reticular and elastic fibers pink.

3.4 Image analysis

Histology samples were examined under light microscope (Eclipse 80i,Nikon Co., Japan) and high resolution (2560x1920 pixel) images were taken with 5 megapixel CCD camera (DS-Fi1, Nikon Co., Japan) either one of the techniques: brightfield light, polarized light and phase contrast condenser, to determine thermal effect of near infrared lasers to the tissue as well as tissue closure abilities. Imaging software (NIS Elements-D, NikonCo., Japan) was used to quantify parameters which give idea on how successfully an incision closed.

Mentioned parameters are: Closure Index (CI) (Figure 9), Thermally Altered Area (TAA) (Figure 10), Granulation Area (GA) and Epidermal Thickness (ET).

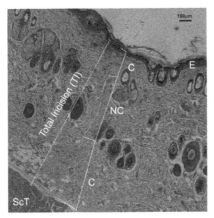

Fig. 9. Computation of Closure Index. C: Closed, NC: Non-closed, E: Epidermis, ScT: Subcutaneous tissue, TI: Total Incision. X4, H&E, Brightfield.

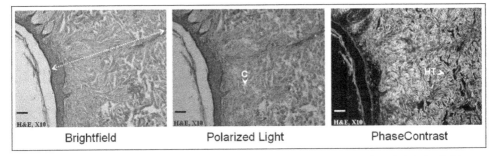

| Brightfield | Polarized Light | PhaseContrast |

Fig. 10. Assessment of thermally altered areas (TAA). Image examined with bright field, polarized light and phase contrast microscopy is same. The bright-field microscopy reveals overall structure soft tissue sample. Incision line was indicated by arrow. The polarized microscopy reveals the collagen molecules (C) that were not affected by heat. Denatured collagen molecules loose their birefringent characteristic. The phase contrast technique reveals the hyperthermal (HT) area. H&E, X10, ScaleBar is 100 μm.

3.5 Tensile test

Dumbbell-shaped stripes, including incision site, was removed. In order to minimize drying, tensile strength tests were performed in between 10 minutes after removal. Samples were placed between single column universal testing machine (LF Plus, Lloyd Instruments, UK) jigs' such a way that grips (TG34, Lloyd Instruments, UK) were as close as possible (2-4 mm). Jigs surfaces were notched like lock-and-key to prevent skin sample slipped through jigs. Each sample was tested at 5 mm/min crosshead velocity. Force (N) and time (s) at break point (this was determined as first opening anywhere along 1 cm long incision) were recorded by software Nexygen Plus.

4. Results and discussion

Laser tissue soldering (LTS) using a diode laser and an indocyanine green (ICG)/albumin solder has been shown to be an effective technique by providing watertight sealant with minimal damage to underlying tissue [82, 83]. Cooper *et al*. [83] tried power densities of 3.2, 8, 15.9, 23.9, 31.8, 47.7, and 63.7 W/cm^2 at constant ICG concentration (2.5 mg/ml). 15.9W/ cm^2 power density was found providing the most controlled heating curve. At a power density of 15.9W/cm^2 or below, increasing ICG concentrations did appear to result in small, but significant, increases in immediate tensile strength as well. 15.92 W/cm^2 power density showed the optimal result in our study in a different aspect: thermal quantitative measurements. Results favoured the 15.92 W/cm^2 power density in laser skin tissue soldering by giving less thermal damage and higher closure rate. On the contrary, the results of the optimal parameter investigation performed by McNally *et al*. Suggest that the strongest repairs are produced with lower irradiances of around 6.4W/cm^2 after approximately 50 seconds exposure time by using a solid protein solder composed of 60% BSA and 0.25 mg/ml ICG on *ex vivo* aorta specimens [74]. Another *in vivo* study performed by Capon *et al*. [43] on mutant OFA Sprague-Dawley rats skin incisions closed over transparent adhesive dressing (Tegaderm, 3M Health Care, Borken, Germany) with the following parameters: 815 nm diode laser, 1.5 W; 3 seconds; spotdiameter: 2 mm; fluence: 145 J/cm^2. The parameters used in this study lead to a slow temperature increase upto a plateau below the critical coagulation temperature (53°C-60°C),avoiding thermal damage [49]. The optimal irradiance (47 W/cm^2) used in this study seems to be very similar to the value reported by Abergel *et al*. [84] (50 W/cm^2). Three times higher laser irradiance was applied for effective skin closure compared to current study without any thermal damage. High irradiation was avoided by use of ICG and use of albumin supplied good apposition for tissue closure.

4.1 Closure index

Laser welding techniques aims immediate healthy tissue closure. So, histological examination in the very early steps of recovery period revealed succes of this technique. (Figure 11).

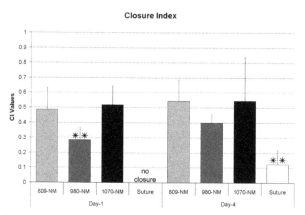

Fig. 11. Closure index values of 809 nm, 980 nm, and 1070 nm laser irradiated and sutured incisions. CI is the ratio of closed segment to total incision (from skin surface to subcutaneous layer). On day 1: immediate closing of laser groups, sutured incisions remained open [85].

4.2 Thermally altered areas

Thermal effects such as hyperthermia and coagulation around laser irradiated area was monitored and quantified (Figure 12).

4.3 Granulation tissue area and epidermal thickness

Granulation tissue had started to form on the 4th postoperative day only in the 809 nm and 1070 nm irradiated samples. Granulation tissue measurement on day 21 showed no statistical difference between experimental groups.

Epidermal thickness (ET) normally starts at a higher value and gradually decreases to a certain level, depending on the degree of the trauma. On day 21, the epidermal thicknesses of all groups' incisions showed no statistical difference and were around the 35-40 μm level (p<0.05). The most impressive results for epidermal thickness were obtained from the 1070 nm laser irradiated samples: there was almost no thickening ((35 μm following the operation and during the 21-day recovery period.

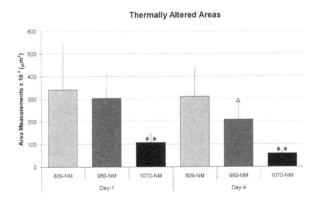

Fig. 12. The thermal hazard of 1070 nm laser welding was minimum on day 1 and on day 4 (p<0.05). Soldering (ICG+BSA) with the 809 nm diode laser was the most thermally hazardous of all the lasers on day 4 (p<0.05). No tissue carbonization was observed in any of the applications [85].

4.4 Tensile tests

Overall tensile strengths are shown in Figure 13. Measurement was not applicable for sutured incisions on day 1 (the incisions had not closed).

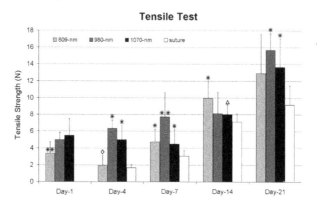

Fig. 13. Mechanical (tensile) test results. On day 1, the tensile strengths of the 809 nm closed incisions were weaker than those of the welded incisions (**) (p<0.05). On day 21, tensile strengths of the laser-welded incisions (980 nm and 1070 nm) were greater than in the control group (*) (p<0.05) [85].

5. Conclusion

This study is a comprehensive comparative research which has accomplished the following tasks:

1. *Comparison of laser welding and suturing methods on a surgical skin incision*

Two closure methodologies served a physical approximation for edges of incision. Laser energy went one step further and managed to seal the gap leakproof. No redness which is a sign of pus formation was observed macroscopically in any of the incision closed with either lasers. Tensile strength gained at the end of the rst week of healing period for suture closed incisions had been already achieved in 24 hours post surgery by laser welding. At the end of 21-Days follow up of recovery period, tensile strengths of laser-welded incisions were greater than in the control group. This shows that laser welding is advantageous over suturing.

2. *Comparison of effects of different wavelengths on skin tissue welding*

Three near infrared wavelengths (809, 980 and 1070 nm) were compared and found to have different welding abilities. Each of these wavelengths should be chosen for specific welding purpose. For instance, any incision on face can be closed by 1070 nm laser for aesthetic appearance whereas strong closure in skin tissue over joints can be obtained by 980 nm laser application.

3. *Establishment of experimental protocols and measurement techniques*

Full-thickness incisions were done through skin tissue. Laser applications were done precisely spot by spot having cooling time intervals in between. Changes in tissue structure were observed by general staining method combined with special microscopy techniques. Macroscopic and microscopic inspections were quantified, resulting in new terminology for future studies: Closure Index, Thermally Altered Area, Granulation Area, and Epidermal Thickness. All the mentioned inspections were performed on particular days at given period of surgical incision healing. Progress of bonding was measured as tensile strength through the same particular days, thus histological findings were evaluated from mechanical point of view or vice versa.

6. Acknowledgment

The authors are grateful for the grant support provided by The Scientific & Technological Research Council of Turkey – TÜBİTAK, contract number, 104M428 and 107E119, and Boğaziçi University Research Fund (contact number: BAP-06M102) The rats were obtained from the Psychobiology Laboratory (Dr.ReşitCanbeyli), BoğaziçiUniversity. The 980 nm diode laser was provided by Dr. İnci Çilesiz, İstanbul Technical University.

7. References

[1] Niemz M H (1996) Laser Tissue Interactions: Fundementals and Applications, Springer. 297 p.

[2] Gürkan Özer A (2008) Comparative Temperature Measurements on Brain and Liver Tissues Using 1070 nm LASER. MSc Dissertation. İstanbul Technical University. 124 p.

[3] Prasad P N (2003) Introduction to Biophotonics. Wiley-Interscience. 593 p.

[4] Yahr W Z, Strully K J (1964) Non-occlusive small arterial anastomosis with neodymium laser. Surg Forum. 15: 224.

[5] McNally-Heintzelman K M, Welch A J (2003) Laser tissue welding. In Tuan V D editor. Biomedical Photonics Handbook. New York: CRC Press LLC. pp. 39(1)-39(45),

[6] Schober R, Ulrich F, Sander T, Durselen H, Hessel S (1986) Laser-induced alteration of collagen substructure allows microsurgical tissue welding. Science. 232: 1421-1422.

[7] Tang J. Godlewski G, Rouy S, Delacretaz G (1997) Morphologic changes in collagen fibers after 830 nm diode laser welding. Lasers Surg Med. 21-4: 438-443

[8] Bass L S, Moazami N, Pocsidio J, (1992) Changes in Type-I collagen following laser welding. Lasers Surg Med. 12: 500-505.

[9] Vale B H, Frenkel A, Trenka-Benthin S (1986) Microsurgical anastomosis of rat carotid arteries with the CO_2 laser. Plast Reconstr Surg. 77: 759-766.

[10] Menovsky T, Beek J F, Van Gemert M J C (1996) Laser tissue welding of dura mater and peripheral nerves: A scanning electron microscopy study. Lasers Surg Med. 19: 152-158.

[11] Godlewski G, Rouy S, Dauzat M (1987) Ultrastructural study of arterial wall repaira after argon laser microanastomosis Lasers Surg Med. 7: 258-262.

[12] Kopchock G E (1988) Mechanism of tissue fusion in argon laser-welded vein artery anastomoses. Lasers Surg Med. 83-89.

[13] Kada O, Shimizu K, (1987) An alternative method of vascular anastomosis by laser: Experimental and clinical study. Lasers Surg Med. 7: 240-248.

[14] Constantinescu M A, Alferi A, Mihalache G, Stuker F, Ducray A, Seiler R W (2007) Effect of laser soldering irradiation on covalent bonds of pure collagen. Lasers Med Sci. 22: 10-14.

[15] Mustoe T A, Pierce G F, Thomason A, Gramates P, Sporn M B, Deuel T F (1987) Accelerated healing of incisional wounds in rats induced by transforming growth factor. Science 237-4820: 1333-1336.

[16] Stadelmann W K, Digenis A G, Tobin G R (1998) Physiology and healing dynamics of chronic cutaneous wounds. Am J Surg (Suppl 2A) 176: 26S-38.

[17] Ross R, Benditt E. P (1962) Wound healing and collagen formation: Fine structure in experimental scurvy. J. Cell Biol. 12: 533-551.

[18] Williamson D, Harding K (2004) Wound healing. Medicine. 32-12: 4-7.

[19] Lindstedt E, Sandblom P (1975) Wound healing in man: Tensile strength of healing wounds in some patient groups. Ann. Surg. 181- 6: 842-846.

[20] Ordman L J, Gillman T (1966) Studies in the healing of cutaneous wounds. Arch Surg. 93-6: 857-882.

[21] Prockop D, (1998) What holds us together? Why do some of us fail apart? What can we do about it? Matrix Biology. 16: 519-528.

[22] Hochberg J, Meyer K M, Marion M D (2009) Suture choice and other methods of skin closure. Surg Clin N Am. 89: 627-641.

[23] Quinn J, Maw J, Ramotar K, Wenckebach G, Wells G (1997) Octylcyanoacrylate tissue adhesive versus suture wound repair in a contaminated wound model. J. Vasc. Surg. 122-1: 69-72.

[24] Quinn J, Wells G, Sutcliffe T, Jarmuske M, Maw J, Stiell I, Johns P (1998) Tissue adhesive versus suture wound repair at 1 year: Randomized clinical trial correlating early, 3-month, and 1-year cosmetic outcome. Annals of Emergency Medicine. 32-6: 645-649.

[25] Leaper D J, Pollock A V, Evans M (2005) Abdominal wound closure: A trial of nylon, polyglycolic acid and steel sutures. British Journal of Surgery. 64-8: 603-606.

[26] Barnett P, Jarman F. C, Goodge J, Silk G, Aickin R (1998) Randomised trial of histoacryl blue tissue adhesive glue versus suturing in the repair of paediatric lacerations. J.Paediatr. Child Health. 34-8: 548-550.

[27] Hollander J, Singer A. J (1998) Application of tissue adhesives: Rapid attainment of proficiency. Academic Emergency Medicine. 5-10: 1012-1017.

[28] Agarwal A, Kumar D, Jacob S, Baid C, Agarwal A, Srinivasan S (2008) Fibrin glue assisted sutureless posterior chamber intraocular lens implantation in eyes with deficient posterior capsules. J. Cat Ref .Surg. 34-9: 1433-1438.

[29] Babetty Z, Sumer A, Altintas S (1998) Knot properties of alternating sliding knots with different patterns in comparison to alternating and simple sliding knots. J.Am Coll Surg. 186-5: 485-489.

[30] Babetty Z, Sumer A, Altintas S, Erguney S, Goksel S (1998) Changes in knot holding capacity of sliding knots in vivo and tissue reaction. Arch Surg. 133: 727-734.

[31] Altman G H, Diaz F, Jakuba C, Calabro T, Horan R L, Chen J, Lu H, Richmond J, Kaplan L (2003) Silk based biomaterials. Biomaterials. 24- 3: 401-416.

[32] Parell G J, Becker G D (2003) Comparison of absorbable with nonabsorbable sutures in closure of facial skin wounds. Arch Facial Plast Surg. 5: 488-490.

[33] Austin P E, Dunn K A, Cofield K E, Brown C K, Wooden W A, Bradfield J. F (1995) Subcuticular sutures and the rate of inflammation in noncontaminated wounds. Ann Emerg Med. 25-3: 328-330.

[34] Moy R L, Waldman B, Hein D W (1992) A review of sutures and suturing techniques. J. Dermatol. Surg.Oncol. 18: 785-795.

[35] Bass L, Treat M (1995) Laser tissue welding: A comprehensive review of current andfuture clinical applications. Lasers Surg Med 17: 315-349.

[36] Barrerias D, Reddy P P, McLorie G A, Bagli D, Khoury A E, Farhat W, Lilge L, Merguerian P A (2000) Lessons learned from laser tissue soldering and fibrin glue pyelo-plasty in an in vivo porcine model. J. Urol. 164-3: 1106.

[37] Wolf J S, Soble J J, Nakada S Y, Rayala H J, Humphrey P A, Clayman R V,. Poppas D P (1997) Comparison of brin glue, laser weld, and mecanical suturing device for laparoscopic closure of ureterotomy in a porcine model. J.Urol. 157-4: 1487.

[38] Zhang L, Kolker A R, Choe E I, Bakshandeh N, Josephson G, Wu F C, Siebert J W, Kasabian A K (1997) Venous microanastomosis with unilink system, sleeve, and suture techniques: A comparative study in the rat. J. Reconst.Microsurg. 13-4: 257.

[39] Pursifull N F, Morey A F (2007) Tissue glues and nonsuturing techniques. Current Opinionin Urology. 17- 6: 396-401.

[40] Durham L H, Willatt D J, Yung M W, Jones I, Stevenson P A, Ramadan M F (1987) A method for preparation of fibrin glue. J. Laryn. Otol. 101: 1182- 1186.

[41] Godlewski G S, (1995) Scanning electron microscopy of microarterial anastomoses with a diode laser: Comparisonwith conventional manual suture. J.Reconstr. Microsurg. 11-1: 33-40.

[42] Hasegawa M, Sakurai T, Matsushita M, Nishikimi N (2001) Comparison of argon-laser welded and sutured repair of inferior vena cava in canine model. Lasers Surg Med. 29-1: 60-68.

[43] Capon A, Souil E, Gauthier B, Sumian C, Bachelet M, Buys B, Polla B S, Mordon S (2001) Laser assisted skin closure (LASC) by using a 815-nm diode-laser system accelerates and improves wound healing. Lasers Surg Med. 28: 168-175.

[44] Chikamatsu E, Sakurai T, Nishikimi N, Yano T, Nimura Y (1995) Comparison of laser vascular welding, interrupted stures, and continuous sutures in growing vascular anastomoses. Lasers Surg Med. 16-1: 32-41.

[45] Çilesiz I F, Thermal Feedback Control During Laser-Assisted Tissue Welding. PhD thesis, The University of Texas at Austin, Texas, USA, 1994.

[46] McNally K M, Dawes J M, Parker A E, Lauto A, Piper J A, Owen E R (1999) Laser-activated solid protein solder for nerve repair: In vitro studies of tensile strength and solder/tissue temperature. Lasers Med Sci. 14: 228-237.

[47] Welch A J, Torres J H, Cheong W F (1989) Laser physics and laser-tissue interaction Texas Heart Inst J. 16: 141-149.

[48] Welch A J, van Gemert M J C, Star W, Wilson B (1995) Optical-Thermal Response of Laser-IrradiatedTissue. New York: Plenum Press.

[49] Welch A J (1984) The thermal response of laser irradiated tissue. IEEE J Quantum Electron. QE-20-12: 1471-1481.

[50] Chen B, Thomsen S L, Thomas R J, Oliver J, Welch A J (2008) Histological and modeling study of skin thermal injury to 2.0 μm laser irradiation. Laser Surg Med. 40: 358-370.

[51] Springer T A, Welch A J (1993) Temperature control during laser vessel welding. Appl Opt. 32-4: 517-525.

[52] Oz M C, Bass L S, Popp H W, Chuck R S, Johnson J P, Trokel S L, M R Treat (1989) In vitro comparison of thulium-holmium-chromium:YAG and argon ion lasers for welding biliary tissue. Lasers Surg Med. 9-3: 245-252.

[53] Menovsky T, Beek J F, van Gemert M J C (1994) CO_2 laser nerve welding: Optimal laser parameters and the use of solders in vitro. Microsurgery. 15-1: 44-49.

[54] Brooks S G, Ashley S, Fisher J, Davies G A, Griths J, Kester R C, Rees M R (1992) Exogenous chromophores for the argon and Nd:YAG lasers: A potential application to laser-tissue interactions. Lasers Surg Med. 12: 294-301.

[55] Fried N M, Walsh T (2000) Laser skin welding: In vivo tensile strength and wound healing results. Lasers Surg Med. 27-1: 55-63.

[56] Optical constants of ice from the ultraviolet to the microwave http://omlc.ogi.edu/spectra/water/data/warren95.dat. Accessed 2012 Feb 16.

[57] Walsh J T, Flotte T J, Deutsch T F (1989) Er:YAG laser ablation of tissue: E⁻ ect of pulse duration and tissue type on thermal damage. Lasers Surg Med. 9: 314-326.

[58] Van Gemert M J C, Welch A J (1989) Clinical use of laser-tissue interactions (1989) IEEE Eng Med Biol Mag. December: 10-13.

[59] Simhon D, Ravid A, Halpern M, Çilesiz I, Brosh T, Kariv N, Leviav A, Katzir A (2001) Laser soldering of rat skin, using fiberoptic temperature controlled system Lasers Surg Med. 29-2: 265-273.

[60] Eyal O, Katzir A (1994) Thermal feedback control techniques for transistor-transistor logic triggered CO2 laser used for irradiation of biological tissue utilizing infrared fiber optic radiometry. Appl Opt. 33- 9: 1751-1754.

[61] Simhon D, Brosh T, Halpern M, Ravid A, Vasilyev T, Kariv N, Katzir A, Nevo Z (2004) Closure of skin incisions in rabbits by laser soldering I: Wound healing pattern. Lasers Surg Med. 35: 1-11.

[62] Brosh T, Simhon D, Halpern M, Ravid A, Vasilyev T, Kariv N, Nevo Z, Katzir A (2004) Closure of skin incisions in rabbits by laser soldering-II: Tensile strength. Lasers Surg Med. 35: 12:17.

[63] Simhon D, Halpern M, Brosh T, Vasilyev T, Ravid A, Tennenbaum T, Nevo Z, Katzir A (2007) Immediate tight sealing of skin incisions using an innovative temperature-controlled laser soldering device in vivo study in porcine skin. Anals of Surgery. 245-2: 206-213.

[64] Lobel B, Eyal O, Kariv N, Katzir A (2000) Temperature controlled CO_2 laser welding of soft tissues: Urinary bladder welding in di⁻ erent animal models (rats, rabbits, and cats). Lasers Surg Med. 26: 4-12.

[65] Çilesiz I, Thomsen S, Welch A J (1997) Controlled temperature tissue fusion: Argon laser welding of rat intestine in vivo, Part one. Lasers Surg Med. 21: 269-277.

[66] Çilesiz I, Thomsen S, Welch A J, Chan E K (1997) Controlled temperature tissue fusion: Ho:YAG laser welding of rat intestine in vivo, Part two. Lasers Surg Med. 21: 278-286.

[67] Spector D, Rabi Y, Vasserman I, Hardy A, Klausner J, Rabau M, Katzir A (2009) In vitro large diameter bowel anastomosis using a temperature controlled laser tissue soldering system and albumin stent. Lasers Surg Med. 41: 504-508.

[68] Lauto A, Trickett R, Malik R, Dawes J, Owen E (1997) Laser activated solid protein bands for peripheral nerve repair: An in vivo study. Laser Surg Med. 21: 134-141.

[69] Hodges D E, McNally K M, Welch A J (2001) Surgical adhesives for laser-assisted wound closure. Journal of Biomedical Optics. 6-4: 427-431.

[70] Chivers R A (2000) In vitro tissue welding using albumin solder: bond strengths and bonding temperatures. Int. J. Adhesion Adhesives. 20: 179-187.

[71] Fujita M, Morimoto Y, Ohmori S, Usami N, Arai T, Maehara T, Kikuchi M (2003) Preliminary study of laser welding for aortic dissection in a porcine model using a diode laser with indocyanine green. Lasers Surg Med. 32: 341:345.

[72] Birch J F, D J Mandley, S L Williams, D R Worrall, P J Trotter, F Wilkinson, P R Bell (2000) Methylene blue based protein solder for vascular anastomoses: An in vitro burst pressure study. Lasers Surg Med. 23: 323-329.

[73] Oz M C, S Prangi, R S Chuck, C C Marboe, L S Bass, R Nowygrod, M R Treat (1990) Tissue soldering by use of indocyanine green dye-enhanced fibrinogen with the near infrared diode laser. J Vasc Surg. 5: 718-725.

[74] McNally K M, B S Sorg E K Chan, A J Welch, J M Dawes, E R Owen (1999) Optimal parameters for laser tissue soldering. Part-I: Tensile strength and scanning electron microscopy analysis. Lasers Surg Med. 24: 319-331.

[75] McNally K M, B S Sorg A J Welch (2000) Novel solid protein solder designs for laser-assisted tissue repair. Lasers Surg Med. 27: 147-157.

[76] Lauto A, M Stoodley, A Avolio, M Sarris, G McKenzie, D D Sampson, L J R Foster (2007) In vitro and in vivo tissue repair with laser-activated chitosan adhesive. Laser Surg Med. 39: 19-27.

[77] Ishihara M, K Nakanishi, K Ono, M Sato, M Kikuci, Y Saito, H Yura, T Matsui, H Hattori, M Uenoyama, A Kurita (2002) Photocrosslinkable chitosan as a dressing for wound occlusion and accelerator in healing process. Biomaterials. 23: 833-840.

[78] Lauto A, D Mawad, L J R Foster (2008) Review adhesive biomaterials for tissue reconstruction. J Chem Technol Biotechnol 83: 464-472.

[79] Geldi C, (2003) Microcontroller based high power 809-nm diode laser design for photodynamic therapy (PDT) applications. Master's Thesis, Boğaziçi University, Istanbul, Turkey.

[80] Geldi C, O Bozkulak, H O Tabakoğlu, S Isci, A Kurt, M Gulsoy (2006) Development of a surgical diode-laser system: Controlling the mode of operation. Photomed and Laser Surg. 24-6: 733-739.

[81] Geldi C, O Bozkulak, H O Tabakoğlu, S Isci, O Kalkanci, A Kurt, M Gülsoy (2004) Microcontroller based surgical diode laser system. Annual national meeting of biomedical engineers. 24: 44-47.

[82] Kirsch A J, M I Miller, T W Hensle, D T Chang, R Shabsigh, C A Olsson, J P Connor (1995) Laser tissue soldering in urinary tract reconstruction: First human experience. Urology. 46-2: 261-266.

[83] Kirsch A J, G M DeVries, D T Chang, C A Olsson, J P Connor, T W Hensle (1996) Hypospadias repair by laser tissue soldering: Intraoperative results and follow up in 30 children. Urology. 48: 616-623.

[84] Abergel R P, R F Lyons, R A White, G Lask, L Y Matsuoka, R M Dwyer, J Uitto, C A Torrance, I L Springfield. Skin closure by Nd:YAG laser welding. J Am Acad Dermatol. 14: 810-814.

[85] Tabakoglu H O, M Gulsoy (2010) In vivo comparison of near infrared lasers for skinwelding. Lasers Med Sci. 25(3): 411-421.

Permissions

The contributors of this book come from diverse backgrounds, making this book a truly international effort. This book will bring forth new frontiers with its revolutionizing research information and detailed analysis of the nascent developments around the world.

We would like to thank Sadık Kara, for lending his expertise to make the book truly unique. He has played a crucial role in the development of this book. Without his invaluable contribution this book wouldn't have been possible. He has made vital efforts to compile up to date information on the varied aspects of this subject to make this book a valuable addition to the collection of many professionals and students.

This book was conceptualized with the vision of imparting up-to-date information and advanced data in this field. To ensure the same, a matchless editorial board was set up. Every individual on the board went through rigorous rounds of assessment to prove their worth. After which they invested a large part of their time researching and compiling the most relevant data for our readers. Conferences and sessions were held from time to time between the editorial board and the contributing authors to present the data in the most comprehensible form. The editorial team has worked tirelessly to provide valuable and valid information to help people across the globe.

Every chapter published in this book has been scrutinized by our experts. Their significance has been extensively debated. The topics covered herein carry significant findings which will fuel the growth of the discipline. They may even be implemented as practical applications or may be referred to as a beginning point for another development. Chapters in this book were first published by InTech; hereby published with permission under the Creative Commons Attribution License or equivalent.

The editorial board has been involved in producing this book since its inception. They have spent rigorous hours researching and exploring the diverse topics which have resulted in the successful publishing of this book. They have passed on their knowledge of decades through this book. To expedite this challenging task, the publisher supported the team at every step. A small team of assistant editors was also appointed to further simplify the editing procedure and attain best results for the readers.

Our editorial team has been hand-picked from every corner of the world. Their multi-ethnicity adds dynamic inputs to the discussions which result in innovative outcomes. These outcomes are then further discussed with the researchers and contributors who give their valuable feedback and opinion regarding the same. The feedback is then collaborated with the researches and they are edited in a comprehensive manner to aid the understanding of the subject.

Apart from the editorial board, the designing team has also invested a significant amount of their time in understanding the subject and creating the most relevant covers. They scrutinized every image to scout for the most suitable representation of the subject and create an appropriate cover for the book.

The publishing team has been involved in this book since its early stages. They were actively engaged in every process, be it collecting the data, connecting with the contributors or procuring relevant information. The team has been an ardent support to the editorial, designing and production team. Their endless efforts to recruit the best for this project, has resulted in the accomplishment of this book. They are a veteran in the field of academics and their pool of knowledge is as vast as their experience in printing. Their expertise and guidance has proved useful at every step. Their uncompromising quality standards have made this book an exceptional effort. Their encouragement from time to time has been an inspiration for everyone.

The publisher and the editorial board hope that this book will prove to be a valuable piece of knowledge for researchers, students, practitioners and scholars across the globe.

List of Contributors

A. Binnaz Hazar Yoruç
Yıldız Technical University, Science and Technology Application and Research Center, Turkey

B. Cem Şener
Marmara University, Faculty of Dentistry, Department of Oral and Maxillofacial Surgery, Turkey

Erhan Akdoğan and M. Hakan Demir
Yıldız Technical University, Turkey

Muhammed Gulyurt
Fatih University, Turkey

Ahmet Koyun and Esma Ahlatcıoğlu
Yıldız Technical University, Science and Technology Application and Research Center

Yeliz Koca İpek
Tunceli University, Faculty of Engineering, Department of Chemical Engineering, Turkey

Mana Sezdi
Istanbul University,Turkey

Haşim Özgür Tabakoğlu
Fatih University, Institute of Biomedical Engineering, İstanbul, Turkey

Ayşen Gürkan Özer
İstanbul Technical University, Faculty of Art and Sciences, İstanbul, Turkey

Printed in the USA
CPSIA information can be obtained
at www.ICGtesting.com
JSHW011422221024
72173JS00004B/634